Big Data Certification Program for Game Changers

Learn. Practice. Get Certified.

ANGEL BERNIZ

International Institute of Executive Careers (IIEC)

Program Contents

Welcome to the Program

Thank you very much for purchasing this book.

First of all, let's connect on LinkedIn and join my **Game Changers** Newsletter on LinkedIn, so that you can get my insights, learnings, and exclusive contents to help you get career success:

https://www.linkedin.com/in/angelberniz

I would love you to share your experience with me after completing the program. You can message me via LinkedIn.

Okay, let's move on!

Certification Program

This book is the foundation for the **IIEC Big Data Certification Program™** available here:

https://thebestmba.org/program/big-data-certification

The Certification Program is available in the online program and will improve your job opportunities and certify your skills gained from this book.

The exams in the online Certification Program are based on the exam questions covered along this book. And once you sign up, you can take your online program whenever you want (no deadline limits). Just sign up today for your Certification Program and certify your skills.

Now, let's start enjoying this journey.

Preface

As we delve further into the 21st century, we are in the midst of a digital revolution, where vast and complex data sets, also known as Big Data, are becoming increasingly important across industries and sectors. The Big Data era brings with it both intricate challenges and transformative opportunities for those who seek to interpret these data sets and harness their potential to shape the future.

The aim of this program is to serve as a comprehensive and accessible resource for data innovators. It is designed to empower participants to effectively navigate, understand, and extract value from Big Data, thereby leading the way in shaping our data-driven future. Drawing from various disciplines such as computer science, statistics, business intelligence, and ethics, this program encourages a well-rounded, interdisciplinary understanding of the Big Data landscape and its implications across various sectors.

Throughout this program, we will provide foundational understanding of core concepts and principles that govern Big Data, and delve into the ethical dilemmas that come with handling vast amounts of information. The program puts forth strategies, best practices, and actionable insights aimed at empowering data innovators to handle Big Data and drive positive change.

As we further explore the role of Big Data in our world, we will scrutinize its potential impacts on areas such as business, healthcare, governance, and more, emphasizing both the enormous opportunities and the complex challenges that lie ahead.

Throughout the program, we will stress the necessity for responsible data handling, cross-disciplinary collaboration, public discourse, and international cooperation to successfully navigate the intricacies of Big Data. By actively participating in shaping the future of Big Data, data innovators can ensure that the data is managed and utilized in ways that align with our collective values, ethical principles, and societal goals.

Let's break down the chapters:

1. **Introduction to Big Data** - This module sets the foundation by defining what big data is, its evolution, significance, and various applications.

2. **Big Data Technologies and Frameworks** - Here, you delve into the technical aspects, exploring key technologies and frameworks, including Hadoop, Spark, NoSQL, and more.

3. **Big Data Storage Solutions** - This module discusses various storage solutions and techniques, which are crucial for managing big data effectively.

4. **Big Data Processing and Analytics** - This module deals with processing methods and how big data can be analyzed, including the application of machine learning and AI, text analytics, and graph analytics.

5. **Big Data Visualization** - This module emphasizes on the importance of data visualization and discusses popular tools, interactive dashboards, and geospatial data visualization.

6. **Data Privacy and Security in Big Data** - Here, the focus is on the critical issues of privacy and security in big data, including regulations, anonymization techniques, encryption, and access control.

7. **Big Data Governance** - This module delves into the essentials of data governance, including data quality and metadata management, and data lineage.

8. **Big Data Challenges and Future Trends** - In this module, you explore the challenges in managing big data and future trends like edge computing, IoT, and quantum computing.

9. **Big Data Case Studies** - The book wraps up with real-life case studies from various industries, providing practical examples of how big data is used.

10. **Building a Career in Big Data** - This module is focused on the human aspect, discussing the roles, and how to grow your skills in order to build a successful big data career.

By presenting this program, it is my ambition that participants will gain a thorough understanding of the Big Data era and its ramifications, along with a sense of optimism and empowerment. As data innovators, you possess a unique opportunity to direct the course of Big Data in ways that resonate with our common goals for a more equitable and sustainable world. Let us together embrace the challenges and seize the opportunities presented by the Big Data era, and together, chart a promising course for the data-driven future.

Acknowledgments

My notes at the university were: "Early adopters are the wealthiest people on earth."

Little did I know how those words would shape my path. As I reflect on my journey, I am overwhelmed with gratitude for the countless individuals who accompanied, guided, and inspired me along the way.

I owe a profound debt of gratitude to my mentors, colleagues, and collaborators in executive education and the technology industry. Their unwavering support and willingness to share their knowledge and perspectives have been invaluable. Their guidance has shaped my career and fueled my passion.

To my students at the International Institute of Executive Careers (IIEC), you have been my constant inspiration. Your curiosity and dedication have made me a better educator and challenged me to refine my ideas. Your insights have significantly contributed to this book.

I also acknowledge the remarkable researchers, professionals, and practitioners who have paved the way for our understanding of the potential of emerging technologies. Their visionary thinking and tireless pursuit of knowledge continue to inspire individuals worldwide, including myself.

I am deeply grateful for the unwavering love, support, and understanding of my friends and family. Your belief in me and willingness to engage with my ideas have given me strength and resilience. This book is dedicated to you with love and gratitude.

As I conclude, I am reminded of the interconnected relationships, experiences, and ideas that have shaped this book. To all who have touched my life, big or small, thank you. Your presence has been truly meaningful. Together, let us continue exploring the evolving landscape of technology, guided by the belief that early adopters can shape the future.

About the Author

Angel Berniz is a seasoned Business and Technology executive, strategist, and advisor with over 20 years of experience in the technology industry. He has held various leadership positions in several companies and has worked with clients in different industries, helping them navigate cloud transformation, innovation, sustainability, emerging technologies, and cybersecurity.

Angel Berniz is also a recognized keynote speaker and author on topics related to business technologies, leadership, and innovation. He has been featured in several conferences, seminars, and publications. He is also a professor and lecturer in international executive programs of several Business Schools.

You can explore Angel Berniz's extensive collection of books available on Amazon:

https://www.amazon.com/author/angelberniz

Module 1. Introduction to Big Data

Introduction to Big Data

Welcome to Module 1 on Introduction to Big Data. In this module, you will embark on a fascinating journey to understand the world of Big Data, an increasingly important concept that is reshaping industries and transforming the way we live and work.

The advent of the digital age has led to an explosion of data, from social media posts and emails to digital images and sensor data from IoT devices. Every day, we are collectively generating quintillions of bytes of data, a volume so large that it defies comprehension. But it's not just the volume that matters; it's what we can do with it. This is the realm of Big Data.

The phenomenon of Big Data goes beyond mere size; it also encapsulates the diversity of data types and the speed at which it is generated and processed. This is often referred to as the 3Vs of Big Data: Volume, Variety, and Velocity. Some experts also add Veracity, or the trustworthiness of the data, and Value, the usefulness of the data, to this definition, emphasizing the complexity and potential of Big Data.

Consider the practical example of a multinational retail company. Every day, this company collects data from various sources: point-of-sale transactions, customer feedback, social media, supply chain logistics, and more. This is Big Data in action, and when harnessed effectively, it can reveal patterns and insights that help the company optimize its operations, enhance its customer service, and innovate its product offerings.

Or think about the healthcare industry. By analyzing vast amounts of data from medical records, wearable devices, and genomics research, healthcare professionals can make more accurate diagnoses, predict disease outbreaks, and personalize medical treatments. This is the power of Big Data.

However, the exploitation of Big Data is not without challenges, including issues related to data privacy, data management, and the need for skilled data professionals. Furthermore, the value of Big Data is not intrinsic; it is the analysis and interpretation of Big Data that unlock its true potential.

In this module, you will explore the core concepts and principles of Big Data, trace its evolution, understand its significance, and learn about its diverse applications. By the end of this module, you will have a solid understanding of what Big Data is, why it matters, and how it is shaping our world.

Get ready to dive into the intriguing world of Big Data. Let's get started!

Defining Big Data

When you first hear the term "Big Data", you might simply think it refers to large volumes of data, and you wouldn't be wrong. However, Big Data encompasses much more than just the sheer size of data. In fact, the term "Big Data" is often defined by the four V's: Volume, Variety, Velocity, and Veracity.

Volume

Volume is perhaps the most obvious aspect. This refers to the massive amounts of data generated every second. To put this into perspective, consider your daily digital activities. From checking your emails and browsing social media feeds to making online transactions and streaming music or videos, you're contributing to the generation of data. Now think about the billions of other people and countless businesses doing the same thing. The amount of data being produced is truly staggering, and it's only going to grow as more people and devices come online.

As an example, consider social media platforms like Facebook or Twitter. Every second, users around the globe are posting status updates, photos, tweets, and videos, creating an enormous volume of data. For instance, you might be running a global brand's social media account. Every day, you're receiving thousands, if not millions, of likes, comments, shares, and direct messages. Analyzing this data can provide valuable insights into your audience's preferences and behaviors, allowing you to tailor your marketing strategy accordingly.

Variety

Variety refers to the many types of data you might have to manage. Gone are the days when data was only text-based and stored neatly in relational databases. Today, data comes in diverse formats: structured data like numbers and dates; unstructured data like text, images, and videos; and semi-structured data like XML and JSON files. You might encounter social media posts, GPS coordinates, transaction records, video feeds, and even sensor data from IoT devices. Each of these data types requires different handling, storage, and analysis techniques.

As an example, think about a smart home setup. Here, you have various devices like thermostats, lights, locks, and appliances, all connected and interacting. Each of these devices produces different types of data. Your smart thermostat records temperature readings and adjustments, your smart lights track on-off times and brightness levels, and your smart lock logs lock-unlock times and user entries. All these varied data types need to be integrated and analyzed to understand and optimize your home's energy use and security.

Velocity

Velocity is about the speed at which data is created, processed, and analyzed. Many applications require real-time or near-real-time data processing. For instance, imagine you're monitoring a live chat support for a product you're selling online. Data on customer inquiries, responses, and resolution times is being generated in real-time. The quicker you can analyze this data and gain insights, the better you can improve your customer support performance.

As an example, let's look at the financial sector, specifically, stock trading. Here, real-time data analysis is critical. Stock prices fluctuate by the second, and trading decisions need to be made almost instantaneously. If you're a trader, the speed at which you receive and analyze market data can mean the difference between a profitable trade and a missed opportunity.

Veracity

Lastly, Veracity is about the trustworthiness of the data. Data is only as valuable as its quality. You may have massive amounts of data, but if it's filled with inaccuracies, inconsistencies, or redundancies, your analyses and insights could be misleading.

Ensuring the data's quality, therefore, is a crucial aspect of working with Big Data.

As an example, consider the healthcare sector. As a healthcare provider, you may use electronic health records (EHRs) to store patient information. This data needs to be extremely accurate and up-to-date. Any inaccuracies or inconsistencies in a patient's medical history, diagnoses, or treatments could lead to incorrect medical decisions and negatively impact patient care.

Now let's look at a practical example that ties all of these aspects together. Let's say you're running an e-commerce website. Every day, you're collecting vast amounts (Volume) of diverse data (Variety) from your users: browsing history, transaction records, user profiles, product reviews, and more. This data is being generated continuously and in real-time (Velocity), and you need to ensure it's accurate and reliable (Veracity) to make informed decisions about product recommendations, pricing strategies, and customer engagement.

In summary, defining Big Data involves acknowledging the scale, complexity, and speed of the data you're working with, as well as the need for quality control. By understanding these aspects, you're better equipped to harness the potential of Big Data in your endeavors.

Evolution of Big Data

To fully appreciate the state of Big Data today, it's important to understand its evolution. The journey of Big Data is a fascinating one, marked by technological advancements, growing data generation, and shifts in how we perceive and use data.

In the early days of computing, data was scarce and expensive to store. You may recall the era of punch cards and magnetic tapes, when data storage was limited and data processing was time-consuming. However, as technology progressed, the cost of storage dropped significantly, and the volume of data began to grow.

By the 1980s and 90s, as computers became more common in businesses and households, data generation started to increase. This was also the era of relational databases and SQL (Structured Query Language). You could now store, retrieve, and manipulate data more efficiently. Yet, most of the data was still structured and relatively small in volume compared to today's standards.

The real data explosion started with the advent of the internet in the mid-90s and early 2000s. Suddenly, data was being created at an unprecedented scale. Every web search, every online transaction, every social media interaction was generating data. And it wasn't just the volume of data that was changing, but also its nature. You were now dealing with more diverse data, including text, images, videos, and more.

This explosion of data posed new challenges and opportunities, and it led to the development of new technologies and frameworks. In the mid-2000s, Hadoop—an open-source framework for storing and processing large datasets—was developed. With Hadoop, you could now store and analyze large volumes of structured and unstructured data across clusters of computers.

The advent of cloud computing in the late 2000s and 2010s further transformed the Big Data landscape. With cloud-based platforms and services, you could store, process, and analyze vast amounts of data without the need for substantial on-premise infrastructure.

Meanwhile, real-time and near-real-time data processing became increasingly important. Technologies like Apache Kafka and Spark emerged to handle the velocity aspect of Big Data. You could now analyze and act on data as it was being generated, opening up new possibilities for real-time decision-making and predictive analytics.

Let's take an example from the retail industry. Imagine you're running an e-commerce website. Initially, you may have used SQL databases to store and manage your customer and transaction data. As your business grew, so did your data. You started using Hadoop to handle your increasing data volume and variety. Then, as customer interactions on your site became more dynamic, you adopted real-time data processing tools to provide personalized recommendations and instant customer service.

In recent years, the focus has also been on machine learning and AI technologies that can automatically analyze and learn from Big Data. For instance, you might use machine learning algorithms to predict customer behavior based on past purchase history and browsing patterns.

Today, Big Data is not just about large volumes of data; it's about making sense of that data to extract valuable insights and make informed decisions. It's about turning data into knowledge and action. As we move forward, we can expect Big Data to continue evolving, driven by advances in technology and an ever-growing thirst for information and insights.

Let's dive into some practical examples that illustrate the evolution of Big Data over the years.

- **Early Stage (1980s-90s):** In the early days of computing, businesses primarily dealt with structured data. For instance, let's say you were running a bookstore in the late

80s. Your data would include information about your inventory, like the title, author, and price of each book. This data would be stored in a relational database, and you would use SQL to manage your inventory, track sales, and generate reports.

- **The Internet Era (Mid-90s to Early 2000s):** With the advent of the internet, the nature and scale of data started changing. Now imagine you took your bookstore online in the late 90s. You would be collecting additional data like user profiles, browsing histories, and online purchase records. Your customer base and, consequently, your data would have grown exponentially. Analyzing this data would provide you with valuable insights into customer behavior and preferences, allowing you to tailor your marketing strategies and improve your services.

- **The Big Data Era (Mid-2000s to Present):** The mid-2000s marked the start of the Big Data era. Continuing with the bookstore example, let's say your online bookstore expanded significantly and started selling a wide range of products. You would now be dealing with massive amounts of diverse data, not just from your website, but also from social media, customer reviews, and more. This is when you would start leveraging Big Data technologies like Hadoop to store and process your data. You might also start using data visualization tools to understand your data better and make more informed decisions.

- **The Real-Time Processing and AI Era (2010s to Present):** As your business continues to grow and customer interactions become more dynamic, real-time data processing becomes crucial. For instance, you might start using a technology like Apache Kafka to analyze customer interactions in real-time and provide personalized product recommendations. Simultaneously, you might start leveraging machine learning algorithms to predict future customer behavior, optimize your inventory, and enhance your marketing strategies.

Through these examples, you can see how the evolution of Big Data has mirrored the evolution of technology and business needs. And as new technologies and methodologies continue to emerge, you can expect the landscape of Big Data to continue evolving.

Importance of Big Data

Understanding the importance of Big Data is crucial for navigating today's data-driven world. Big Data is more than just a buzzword; it's a fundamental shift in how you can gather, analyze, and leverage information. Its significance spans various domains, offering you the opportunity to unlock insights, improve decision-making, and drive innovation.

Firstly, Big Data can provide you with deeper and broader insights. The vast volume and variety of data available mean you can analyze complex patterns, trends, and relationships that were previously hidden or inaccessible. For example, if you're running an e-commerce website, analyzing your customer data can reveal insights about their shopping habits, preferences, and behaviors, allowing you to personalize their shopping experience and improve customer retention.

Secondly, Big Data can enhance your decision-making capabilities. With access to real-time data and sophisticated analytics tools, you can make informed decisions swiftly and accurately. In the context of a healthcare provider, you could use real-time patient data to make quick and accurate diagnoses, monitor patient health effectively, and provide personalized treatment plans.

Moreover, Big Data can drive innovation and competitiveness. By analyzing and learning from your data, you can identify opportunities for new products, services, or business models. Suppose you're a tech entrepreneur. You could use Big Data to identify gaps in the market, develop innovative solutions, and stay ahead of your competition.

Additionally, Big Data can improve operational efficiency. By analyzing your operational data, you can identify bottlenecks, inefficiencies, and areas for improvement. If you're managing a supply chain, for instance, you could use Big Data to optimize your logistics, reduce costs, and improve delivery times.

However, harnessing the power of Big Data comes with its own set of challenges. Data privacy, data security, data quality, and the need for skilled data professionals are key considerations. Therefore, while leveraging Big Data, it's essential to have robust data governance and management practices in place.

Let's consider some practical examples of how Big Data is important across different sectors.

- **Retail:** As a retailer, you can use Big Data to understand your customers better, personalize their shopping experience, optimize your inventory, and improve your marketing strategies. By analyzing customer feedback, purchase histories, and browsing patterns, you can predict future trends, tailor your offerings, and boost sales.

- **Healthcare:** In the healthcare sector, Big Data can transform patient care. By analyzing patient records, wearable device data, and genomic data, you can provide personalized treatments, predict disease outbreaks, and enhance patient outcomes.

- **Finance:** If you're working in the financial sector, Big Data can help you detect fraudulent activities, manage risks, and make informed investment decisions. By analyzing transaction data, market data, and social media data, you can spot unusual patterns, predict market trends, and provide personalized financial advice.

- **Manufacturing:** In the manufacturing industry, you can use Big Data to optimize your production processes, improve quality control, and reduce downtime. By analyzing machine data, process data, and quality data, you can predict equipment failures, streamline your operations, and improve product quality.

In summary, Big Data is reshaping the way you do business, make decisions, and interact with the world. It's not just about having more data; it's about what you can do with it.

Big Data Applications and Use Cases

Big Data applications and use cases span a multitude of sectors and industries, each harnessing the power of data in unique ways. Let's explore some key applications and how they can directly impact your work or daily life.

Starting with healthcare, one of the most compelling applications of Big Data is in personalized medicine. By analyzing your medical history, genetic profile, and lifestyle factors, healthcare providers can design personalized treatment plans, enhancing effectiveness and reducing potential side effects. Similarly, Big Data is fueling advancements in predictive analytics, allowing for early detection of diseases and conditions, potentially even before symptoms appear.

In the financial sector, Big Data is revolutionizing risk management and fraud detection. By analyzing your transaction patterns along with those of millions of others, banks and credit card companies can identify suspicious activity almost instantly. This real-time analysis of Big Data allows them to alert you promptly, safeguarding your financial assets.

If you're in retail or e-commerce, Big Data can transform the way you understand your customers. By analyzing customer behavior, purchase history, and feedback, you can personalize your offerings and enhance your customers' shopping experience. Moreover, predictive analytics can help you anticipate future trends and demands, giving you a competitive edge in your market.

The power of Big Data is also evident in the realm of social media. By analyzing your likes, shares, and posts, social media platforms can personalize your feed and recommend content that aligns with your interests. This personalization not only enhances your user experience but also makes advertising on these platforms more targeted and effective.

In the field of transportation and logistics, Big Data can dramatically improve efficiency. By analyzing traffic data, vehicle data, and weather data, you can optimize your routes, reduce fuel consumption, and ensure timely deliveries.

For a more detailed look at these applications, consider the following practical examples:

- **Personalized Medicine:** Imagine you're a patient with a specific chronic condition. Traditional treatment methods may have a varying degree of success. However, with Big Data, your doctor can analyze vast amounts of data from similar cases, combine it with your personal medical history and genetic information, and devise a treatment plan tailored specifically for you. This individualized approach increases the chances of successful treatment and reduces potential side effects.

- **Fraud Detection:** Let's say you're a credit card user. One day, a series of suspicious transactions appear on your account. Thanks to Big Data analytics, your bank detects these anomalies in real-time, alerts you, and blocks your card to prevent further fraudulent activity. Without Big Data, these fraudulent transactions might have gone unnoticed until you reviewed your monthly statement.

- **Customized Marketing:** Suppose you own an online clothing store. By using Big Data analytics, you can analyze your customers' purchase histories, browse times, and even abandoned carts. This analysis allows you to personalize your marketing campaigns for each customer, offering them deals on the items they are most likely to purchase.

- **Traffic Optimization:** Imagine you're managing a fleet of delivery trucks. By using Big Data, you can analyze real-time traffic data, weather forecasts, and historical traffic patterns to determine the most efficient routes for your drivers. This optimization can save time, reduce fuel consumption, and ensure timely deliveries.

Understanding the broad applications and use cases of Big Data across industries will allow you to better appreciate its potential and the value it can bring to your own endeavors. Whether you're a business owner, a healthcare provider, a consumer, or a city planner, Big Data can provide insights and benefits that were unimaginable just a few years ago.

Module 1 Exam Questions

1. Which of the following best describes 'Big Data'?

a) Large amounts of structured data

b) Small amounts of unstructured data

c) Large amounts of varied data

d) Small amounts of consistent data

2. Which of the following is NOT one of the 4Vs of Big Data?

a) Velocity

b) Variety

c) Volume

d) Visibility

3. Which V of Big Data refers to the speed at which data is generated and processed?

a) Velocity

b) Variety

c) Volume

d) Veracity

4. Which era marked the beginning of Big Data?

a) The Internet Era

b) The Big Data Era

c) The Early Stage

d) The Real-Time Processing and AI Era

5. What is a practical example of the use of Big Data in the healthcare sector?

a) Fraud detection

b) Personalized medicine

c) Traffic optimization

d) Personalized marketing

6. What does Big Data analytics in the financial sector help to prevent?

a) Disease outbreaks

b) Traffic congestion

c) Fraudulent activities

d) Inventory shortages

7. In the context of Big Data, what does 'variety' refer to?

a) Different types of data

b) Large amounts of data

c) Speed of data processing

d) Trustworthiness of data

8. Which of these is a challenge in harnessing the power of Big Data?

a) Data security

b) Data variety

c) Data velocity

d) Data volume

9. How can Big Data improve customer experience in the retail sector?

a) By optimizing routes

b) By diagnosing diseases

c) By personalizing offerings

d) By preventing fraud

10. How does social media use Big Data to enhance user experience?

a) By optimizing routes

b) By diagnosing diseases

c) By personalizing feeds

d) By preventing fraud

11. How can Big Data improve operational efficiency in the supply chain?

a) By optimizing routes

b) By diagnosing diseases

c) By personalizing feeds

d) By preventing fraud

12. What role does Big Data play in risk management in the financial sector?

a) It helps in fraud detection

b) It helps in predicting traffic patterns

c) It helps in diagnosing diseases

d) It helps in personalizing marketing campaigns

13. Which of these sectors can benefit from Big Data?

a) Healthcare

b) Finance

c) Retail

d) All of the above

14. How can Big Data analytics benefit an online clothing store?

a) By predicting disease outbreaks

b) By predicting future trends and demands

c) By detecting fraudulent activities

d) By optimizing traffic routes

15. How can Big Data help in early detection of diseases?

a) By analyzing medical history and genetic profile

b) By analyzing transaction patterns

c) By analyzing customer behavior

d) By analyzing traffic data

16. Which of the following is NOT a use case of Big Data?

a) Personalized Medicine

b) Fraud Detection

c) Traffic Optimization

d) All of the above are use cases of Big Data

17. In the era of Big Data, what can healthcare providers design by analyzing medical history, genetic profile, and lifestyle factors?

a) Personalized shopping experience

b) Personalized treatment plans

c) Personalized marketing campaigns

d) Personalized travel routes

18. In the context of Big Data, what does 'veracity' refer to?

a) Large amounts of data

b) Different types of data

c) Trustworthiness of data

d) Speed of data processing

19. Which of these sectors uses Big Data to optimize routes, reduce fuel consumption, and ensure timely deliveries?

a) Healthcare

b) Finance

c) Retail

d) Transportation and logistics

20. How does Big Data analysis help in enhancing the user experience on social media platforms?

a) By diagnosing diseases

b) By predicting traffic patterns

c) By personalizing feeds

d) By preventing fraud

Answers:

1. c
2. d
3. a
4. b
5. b
6. c
7. a
8. a
9. c
10. c
11. a
12. a
13. d
14. b
15. a
16. d
17. b
18. c
19. d
20. c

Module 2. Big Data Technologies and Frameworks

Introduction to Big Data Technologies and Frameworks

Welcome to Module 2 on Big Data Technologies and Frameworks. In this module, you will delve into the heart of big data, exploring the key technologies and frameworks that form the backbone of any big data project.

Understanding the technologies and frameworks that power big data is fundamental to grasping the capabilities and limitations of various tools and systems. As we navigate through this module, you will gain a deeper understanding of how these technologies work, their application, and their significance in different contexts.

We will embark on a journey that starts with the Hadoop ecosystem, an open-source framework that revolutionized the big data industry, providing a way to store and process large volumes of data across clusters of computers. We'll then proceed to Spark, a powerful processing engine designed for speed and ease of use.

From there, we'll explore NoSQL databases, which challenge the traditional database models and offer more flexibility and scalability for handling diverse data types. Next, we'll venture into the realm of data warehouses and data lakes, contrasting their respective features and uses.

Finally, we'll conclude the module with a look at stream processing, a method of analyzing and acting on real-time data, which is becoming increasingly crucial in today's fast-paced digital world.

Each of these topics will not only be discussed theoretically, but we will also look at practical examples to highlight their real-world applications. For instance, we might examine how Hadoop's distributed file system enables a multinational corporation to analyze petabytes of data seamlessly, or how a tech startup might use Spark to process live user interactions for a personalized recommendation system.

In understanding these technologies and frameworks, you will be better equipped to design and implement effective big data strategies, select the most appropriate tools for your specific needs, and navigate the dynamic landscape of big data.

By the end of this module, you will have a comprehensive understanding of the various technologies and frameworks that underpin big data. You will be equipped with knowledge that will serve as a solid foundation as we delve deeper into the complexities and intricacies of big data in the subsequent modules.

So, let's embark on this fascinating journey to unravel the technologies and frameworks powering the world of big data!

Hadoop Ecosystem

As you delve deeper into the world of big data, it's impossible to overlook the significance of the Hadoop ecosystem. The Hadoop ecosystem is a comprehensive suite of open-source tools and utilities that leverage the power of distributed computing to handle big data. It was born out of the necessity to process an enormous volume of data at a scale that traditional data processing systems could not handle.

Let's begin by understanding the core components of the Hadoop ecosystem:

- **Hadoop Distributed File System (HDFS):** HDFS is the backbone of Hadoop and the primary storage system used by it. It's a distributed file system that provides high throughput access to application data and is designed to span large clusters of commodity servers. What makes HDFS unique is its ability to store vast amounts of data across multiple nodes, thus offering you the capability to scale your storage needs horizontally.

- **MapReduce:** MapReduce is a programming model that allows you to process large data sets in parallel across a Hadoop cluster. The Map phase performs sorting and filtering, while the Reduce phase provides a summary of this data. The magic of MapReduce lies in its simplicity: you only need to think about your computation in terms of individual records, and the framework handles everything else.

- **YARN (Yet Another Resource Negotiator):** Introduced in Hadoop 2.0, YARN is a resource management layer that allocates system resources to various running applications and schedules tasks to be executed on different cluster nodes. YARN is crucial as it allows multiple data processing engines such as interactive SQL, real-time streaming, and machine learning to handle data stored in a single platform – Hadoop.

Beyond these core components, the Hadoop ecosystem includes a variety of other tools designed to enrich and ease your work with big data:

- **Hive and Pig:** These tools simplify the process of writing complex MapReduce tasks. Hive provides a SQL-like interface to query data stored in a Hadoop cluster, making it easier if you're familiar with SQL. Pig, on the other hand, is a high-level platform that uses a language called Pig Latin, designed for expressing data transformations in a more straightforward manner than Java-based MapReduce.

- **HBase:** HBase is a NoSQL database that runs on top of HDFS. It is designed to provide real-time read/write access to your big data. With its capability to handle billions of rows and millions of columns, HBase empowers you to run applications that require massive datasets in real-time.

- **Sqoop and Flume:** These tools facilitate data ingestion. Sqoop is designed to transfer data between Hadoop and relational databases, while Flume helps you collect, aggregate, and move a large amount of log data into HDFS.

- **ZooKeeper:** It's a centralized service for maintaining configuration information, naming, and providing distributed synchronization. ZooKeeper helps in avoiding the potential pitfalls of distributed computing like race conditions and deadlocks.

- **Oozie:** Oozie is a server-based workflow scheduling system to manage Hadoop jobs. If you have a series of jobs to be run in a specific order, Oozie is your go-to tool.

- **Mahout and Spark MLlib:** These are machine learning libraries. Mahout integrates with Hadoop to build scalable machine learning algorithms, while Spark MLlib is Apache Spark's scalable machine learning library.

Practical scenario

Now, let's imagine a practical scenario to understand how these components can come into play. Suppose you're working on a project where you need to analyze the sentiments of tweets about a particular product in real-time.

Your data source is the Twitter API, which sends a massive stream of tweets. You can use Flume to ingest this data in real-time into your Hadoop cluster. As tweets arrive, they are stored in HDFS, distributed across your cluster.

Given the nature of the data, you might choose to store the tweets in HBase. HBase, with its low-latency access, is perfect for storing semi-structured data like tweets and allowing real-time read/write operations.

Now, the analysis part comes in. You want to understand the sentiment of each tweet - this is a classic machine learning problem. Here, you could use Mahout or Spark MLlib to train a sentiment analysis model. Perhaps you've decided to use the Bag of Words model for your analysis, which considers the frequency of words in analyzing sentiment.

For this, you first need to convert your tweets into a numerical form that the model can understand. Here's where MapReduce comes in. You write a MapReduce job in which the Map function tokenizes each tweet (splits it into individual words) and the Reduce function counts the frequency of each word.

Once your data is transformed, you feed it into your machine learning model. After your model is trained, you can use it to analyze new tweets in real-time.

As your project grows, you may need to manage multiple tasks: data ingestion, MapReduce tasks, machine learning tasks, etc.

Oozie comes in handy here, allowing you to schedule and manage all these tasks efficiently.

Meanwhile, ZooKeeper will help you maintain the overall application coordination, ensuring that all parts of your distributed system interact harmoniously and data reliability is maintained across the nodes.

Finally, to query your processed data or perform additional analysis, you can use Hive. Hive's SQL-like interface would allow you to interact with your data using a familiar SQL syntax, abstracting away the complexities of the underlying MapReduce operations.

And there you have it - a practical example of using the Hadoop ecosystem. The beauty of the Hadoop ecosystem lies in its modularity. You can pick and choose the components that best fit your project's requirements, and they all work together seamlessly, allowing you to tackle big data challenges effectively.

As you progress with the Hadoop ecosystem, you'll realize it's not just a set of tools but a complete environment for your big data journey, facilitating storage, processing, analysis, and more. The more you explore, the more you'll discover how these elements can empower your big data solutions.

Spark

As we dive deeper into the world of big data technologies, it's essential to acquaint ourselves with Apache Spark. This powerful open-source, distributed computing system has become a prominent tool in the big data landscape due to its speed, ease of use, and versatility. It's designed to cover a wide range of workloads such as batch applications, iterative algorithms, interactive queries, and streaming.

Understanding Spark

At the core of Apache Spark is its ability to process data in memory. This in-memory processing capability allows Spark to deliver speed that can be up to 100 times faster for certain applications than traditional Hadoop MapReduce. This speed, combined with its ability to compute across multiple nodes, makes Spark a compelling tool when dealing with large datasets.

Spark consists of multiple components that provide flexibility in how you interact with data:

- **Spark Core:** This is the foundation of the overall project. It provides distributed task dispatching, scheduling, and basic I/O functionalities. Spark Core is also home to the API that defines resilient distributed datasets (RDDs), which are Spark's primary data abstraction. RDDs can be created from data stored in HDFS, local file systems, and more.

  Example: Let's say you are analyzing a dataset containing information about various cities around the world, including population size, GDP, and geographical data. Using Spark Core, you could create an RDD from this data and perform transformations and actions on it. For instance, you might want to filter out cities with populations below a certain threshold or calculate the average GDP across all cities.

- **Spark SQL:** This module allows you to interact with structured and semi-structured data using SQL or a DataFrame API. You can connect it with your Hive data warehouse and run SQL queries on data stored there.

  Example: Suppose you have a structured dataset of movie ratings and you want to find the average rating for each movie. You can load this data as a DataFrame and use Spark SQL to run SQL queries directly on it. For instance, you might run a query like "SELECT movie_id, AVG(rating) FROM movie_ratings GROUP BY movie_id" to get the average rating for each movie.

- **Spark Streaming:** With this component, you can process live data streams in real time. Whether data is ingested from Kafka, Flume, or a simple TCP socket, Spark Streaming can handle it.

  Example: Let's say you're monitoring Twitter for mentions of a particular hashtag in real time. Using Spark Streaming, you can ingest the live stream of tweets, filter for your hashtag, and perform any necessary analysis or data processing on the fly - for example, you might want to track the sentiment of these tweets over time.

- **MLlib:** This is Spark's machine learning library. It consists of common learning algorithms and utilities, including classification, regression, clustering, collaborative filtering, and others.

  Example Suppose you're an e-commerce company and you have a large dataset of past customer transactions. You want to create a recommendation system to suggest products to customers based on their past purchases. Using MLlib, you can employ collaborative filtering techniques (like Alternating Least Squares) to build this recommendation system.

- **GraphX:** GraphX is Spark's API for graph computation, allowing you to create a directed graph with arbitrary properties attached to each vertex and edge.

  Example: Let's imagine you have a dataset of social media connections and you want to analyze the network. Each person can be a vertex, and each relationship can be an edge. Using GraphX, you can easily create this graph and perform analyses to find the most influential people in the network, identify communities within the network, and more.

Practical scenario

Now, let's bring this to life with a practical scenario.

Imagine you're working with a large dataset of customer reviews for a range of products. You need to analyze these reviews to identify trends and gain insights into customer sentiment towards the products.

Data Ingestion and Processing:

You could start by loading your data into Spark. Your data might be in a structured format like CSV or a semi-structured format like JSON, residing in HDFS or a local file system. With Spark, you can easily load this data as an RDD or a DataFrame, depending on your needs.

Data Analysis:

Once your data is loaded, you might want to clean it up and format it for analysis. With Spark SQL, you can query your data just like you would with a traditional SQL database. You can filter reviews by product, by date, or any other parameter. The flexibility and power of SQL is available to you right within Spark.

Machine Learning:

With your data now structured and cleaned, suppose you want to perform sentiment analysis on the reviews. For this, you could use Spark's MLlib. You might choose to train a logistic regression model, for example, using a Bag of Words approach to convert your text data into a format the model can understand. MLlib makes this process straightforward, and since it's running on Spark, you can train your model on a much larger dataset than would be possible with a single machine.

Real-time Analysis:

Now, suppose in addition to the historical reviews, new reviews are coming in continuously, and you want to analyze them in real-time. This is where Spark Streaming shines. It allows you to ingest the live data stream and perform operations on it just like you would with an RDD. You can calculate rolling metrics, such as a moving average rating for each product.

Graph Analysis:

Finally, maybe you want to analyze the relationships between products based on the reviews. You might want to see which products are often mentioned together, indicating that they might be complementary. GraphX allows you to do this. You can create a graph where each vertex represents a product and each edge represents a connection between products. The connections could be weighted based on the frequency of mentions in the same review.

By using Spark's GraphX, you could uncover interesting insights. Maybe you find that two products are frequently bought together, suggesting a potential for bundling them in a promotion. Or you might identify a product that is a central node in the graph, indicating it is frequently bought with many other products.

Conclusion

Overall, Spark's robust tools and capacity for large-scale data processing are essential for big data handling. The power of Spark lies in its versatility and its ability to scale, allowing you to tackle a wide range of big data challenges. The more you delve into Spark, the more ways you'll find to leverage its capabilities to extract meaningful insights from your data.

NoSQL Databases

As we delve deeper into big data technologies, we encounter NoSQL databases - a type of database design that provides flexible, scalable, and high-performance solutions for managing and querying unstructured data. NoSQL, which stands for "Not Only SQL", emerged as a response to the limitations of traditional relational databases (SQL databases) in handling the volume, variety, and velocity of big data.

Understanding NoSQL Databases

NoSQL databases are designed for distributed data stores where very large scale of data storing needs are required. They are especially useful when an enormous amount of data needs to be stored and accessed, but without a well-defined schema. NoSQL databases can store relationship data - they're just designed differently than relational databases.

There are four main types of NoSQL databases, each with its specific use-case:

- **Key-Value Stores:** These are the simplest type of NoSQL databases. Every single item in the database is stored as an attribute name (or key), together with its value. Examples of key-value stores are Riak and Voldemort. These are highly performant and often used for caching and storing session data.

- **Document Databases:** These pair each key with a complex data structure known as a document. Documents can contain many different key-value pairs, or key-array pairs, or even nested documents. MongoDB and CouchDB are examples of document databases. These are used for content management systems, catalogues, and user profiles.

- **Wide-Column Stores:** Instead of rows and columns, these databases use tables, columns, and rows, but unlike a relational database, names and format of the columns can vary from row to row across the table. They are designed to store data tables as sections of columns of data, rather than as rows of data. Examples include Cassandra and HBase. These databases are used for analyzing large datasets and for heavy write loads.

- **Graph Databases:** These are used to store data whose relations are best represented in a graph. Data is saved in graph structures with nodes (entities), properties (information about the entities), and lines (connections between the entities). Examples include Neo4j and OrientDB. These are used for social networks, fraud detection, and recommendation systems.

NoSQL databases offer several benefits:

- **Scalability:** NoSQL databases are horizontally scalable, which means that they can handle more traffic simply by adding more servers to the database. NoSQL databases are a great fit for big data scenarios and real-time web applications.

- **Flexibility:** NoSQL databases allow for flexible schema design. This means that unlike relational databases, you can store different types of data in different ways. This can be particularly useful when the shape of your data doesn't fit well into a tabular format.

- **Speed:** NoSQL databases have been optimized for specific data models (such as document, graph, columnar) that enable higher performance than trying to accomplish similar functionality with relational databases.

NoSQL Databases Types

As we delve into the practical aspects of NoSQL databases, it's essential to remember that NoSQL is not a specific kind of database, but a categorization of several database types that do not follow the traditional relational model. The prominent types include document databases, key-value stores, wide-column stores, and graph databases. Let's walk through a few practical examples, with a focus on MongoDB, Redis, Cassandra, and Neo4j.

MongoDB

As a document database, MongoDB excels at handling semi-structured data. In practical terms, it means you can store data in a format similar to JSON (called BSON in MongoDB), which is flexible and adaptable. This adaptability makes MongoDB a good fit for scenarios where your data structure might change over time. For example, if you are developing a content management system where each piece of content may have different attributes, MongoDB allows you to store each piece as a document with different fields, rather than trying to fit it into a pre-defined schema.

Redis

Redis, on the other hand, is a key-value database. It is renowned for its speed and efficiency, storing data in memory for quick access. A practical example of using Redis could be session management for a web application. Suppose you want to store user session information for your website quickly and retrieve it instantly. In that case, Redis provides an efficient way to store this data as key-value pairs, where the key could be a session ID, and the value is the session data.

Cassandra

Cassandra is a wide-column store, built for scalability and high availability. It's a good fit for managing large amounts of data across many commodity servers. A real-world example of its application is in the Internet of Things (IoT) space. Suppose you have an application that collects data from numerous devices and sensors. This data is vast and writes-heavy, making Cassandra an ideal choice for such an application. Its distributed architecture allows the data to be spread across many nodes, facilitating quick write times and resilience.

Neo4j

Neo4j is a graph database, designed for storing data whose relations are as important as the data itself. It excels at handling complex, interconnected data, making it a good choice for social networking applications, recommendation systems, or any system where understanding the relationships between entities is key. For example, in a social network, you can represent each user as a node and each interaction as an edge. With Neo4j, querying these relationships is intuitive and efficient.

Practical scenario

Let's now understand how you can practically use NoSQL databases.

Suppose you are building an e-commerce platform. You have a range of different data types to deal with, including user profiles, product catalogs, and customer behavior data.

- **User Profiles:** Here, a document database like MongoDB would be a great fit. You can create a document for each user, with key-value pairs for each user attribute. For example, a user document might include keys for username, password, email, and purchase history, each paired with its respective value. The flexible schema allows you to add new attributes as needed.

- **Product Catalog:** A product catalog typically includes a wide variety of items, each with different attributes. Some products might have size and color options, while others might have weight and manufacturer details. A document database like MongoDB would be a great fit for this scenario as well, as it allows you to easily store this diverse range of product attributes.

- **Customer Behavior Data:** Suppose you want to analyze the behavior of your customers - for example, you might want to build a recommendation system based on the browsing and purchase history of your users. This is a classic use case for a graph database like Neo4j. Each user can be represented as a node in the graph, with edges representing their interactions with products. This would allow you to easily analyze the relationships between users and products and make accurate recommendations.

- **Managing High-Traffic:** As your e-commerce platform grows and you start getting a high volume of read and write requests, you might need to scale your databases. This is where the scalability of NoSQL databases really shines. If you're using a key-value store like Riak for session data or a wide-column store like Cassandra for storing large amounts of data, you can easily add more servers to handle the increased load.

In conclusion, NoSQL databases provide a flexible, scalable, and high-performance solution for managing and querying big data. They can handle a wide variety of data types and are designed to scale with your needs, making them a crucial tool in your big data toolkit. As you continue your big data journey, you'll find many opportunities to use NoSQL databases to store, analyze, and gain insights from your data.

Data Warehouses and Data Lakes

As we progress in our exploration of big data technologies, it's essential to discuss two significant concepts: Data Warehouses and Data Lakes. Both of these structures serve as primary repositories for raw and transformed data, but they cater to different use-cases and data requirements.

Data Warehouses

A data warehouse is a large collection of business data used to help an organization make decisions. It is a blend of technologies and components which allows the strategic use of data. A data warehouse stores data in an organized manner, structured into tables, rows, and columns, similar to a relational database. However, a data warehouse is more than just a database; it includes infrastructure, storage, and tools to extract valuable insights from data.

Data warehouses are optimized for analyzing structured data for business intelligence activities, including analytics, reporting, and dashboarding. The data stored in a warehouse is typically cleaned, transformed, and cataloged. It is designed so that it's easy to understand the data, create reports, and perform complex queries.

One popular example of a data warehouse solution is Amazon Redshift. It's a fully managed, petabyte-scale data warehouse service in the cloud that allows you to analyze data using your existing business intelligence tools. Other examples include Google BigQuery, Snowflake, and Microsoft Azure Synapse Analytics.

Data Lakes

On the other hand, a data lake is a vast pool of raw data, the purpose for which is not yet defined. It allows you to store all your structured and unstructured data at any scale. You can store your data as-is, without having to first structure the data, and run different types of analytics—from dashboards and visualizations to big data processing, real-time analytics, and machine learning to guide better decisions.

Data lakes are designed to be a more flexible solution than data warehouses. They can store any data, regardless of its source or format, including raw copies of source system data, sensor data, social media data, clickstream data, server logs, and more. This flexibility means that you can adjust your analysis approach based on business needs rather than being constrained by the structure of your data.

Apache Hadoop, an open-source software framework for distributed storage and processing of large datasets, is often used to implement data lakes due to its flexibility and scalability. Another popular data lake solution is Amazon S3, which allows you to store and retrieve any amount of data, at any time, from anywhere on the web.

Practical Examples

Let's consider some practical scenarios to understand better how and when to use a data warehouse or a data lake.

Suppose you're running an e-commerce business. You collect a vast amount of data from various sources like your website, mobile app, social media, customer emails, and more.

Data Warehouse Scenario:

For your monthly business report, you need to analyze your sales data. This data is structured and comes from your transactional database. You need to join it with data from your customer database to calculate metrics like sales per customer or sales by region. In this case, a data warehouse would be the ideal solution. You can extract data from your transaction and customer databases, transform it into a suitable format, and load it into the warehouse. From there, you can use SQL queries or a business intelligence tool to analyze the data and generate your report.

Data Lake Scenario:

However, suppose you also want to analyze customer behavior on your website to understand better how your customers interact with your products. You have data from your website logs, which includes information like the sequence of pages visited by a user, the time spent on each page, and the links clicked. This data is semi-structured and doesn't fit well into a tabular format. In this case, a data lake would be the ideal solution. You can store your raw web log data in the data lake without needing to define a schema upfront. From there, you can use big data processing tools like Apache Hadoop or Spark to analyze the data and extract insights about your customers' behavior.

It's also worth noting that a data warehouse and a data lake are not mutually exclusive. Many businesses use both in conjunction. For instance, you could use a data lake to store all of your raw data and a data warehouse for storing the subset of your data that is cleaned, structured, and ready for analysis.

In fact, an emerging concept is that of a "data lakehouse," which seeks to combine the best features of data lakes and data warehouses into a single, unified architecture. A data lakehouse maintains the vast storage and flexibility advantages of a data lake for ingesting raw data while also incorporating the data management features of a data warehouse for handling structured data. Databricks is one organization leading the way in this space, although the concept is still relatively new.

In conclusion, both data warehouses and data lakes play crucial roles in big data architecture. Understanding their strengths, weaknesses, and ideal use cases is key to designing a data strategy that meets your needs. Whether you're dealing with structured or unstructured data, need to run complex analyses or simply store raw data for future use, there's a solution that's right for you.

Stream Processing

We've already covered substantial ground in the realm of big data technologies, and now we're going to turn our attention to a critical capability: stream processing.

Stream processing is a computational paradigm that allows you to process data in real-time as it arrives, rather than storing it and processing it later. This contrasts with batch processing, which involves collecting data over a period of time and processing it all at once. Stream processing is essential for use-cases where immediate insights from data are necessary, such as real-time analytics, anomaly detection, or event-driven applications.

What is Stream Processing?

Stream processing involves ingesting a continuous flow of data and applying computations to it in real-time. This data could be anything from tweets, blog posts, financial transactions, IoT sensor readings, website clicks, or logs from a web server. The computations can be as simple as counting the number of occurrences of a specific event or as complex as machine learning algorithms for real-time predictions.

There are two main components in a stream processing system: the data stream and the stream processing engine. The data stream is a sequence of data records that are produced over time. The stream processing engine is the software that ingests the data stream and applies the computations. The engine might also handle tasks like data partitioning, fault tolerance, and state management.

Stream Processing Technologies

Several technologies are used for stream processing, including Apache Kafka, Apache Flink, Apache Storm, and Apache Samza.

Apache Kafka is a distributed streaming platform that's used for building real-time data pipelines and streaming applications. Kafka allows you to publish and subscribe to streams of records, similar to a message queue or enterprise messaging system.

Apache Flink, Storm, and Samza are all stream processing frameworks that can be used in conjunction with Kafka. They provide the computational logic that is applied to the data stream. These frameworks allow you to implement complex stream processing tasks like event time processing, windowing, and state management.

Practical escenario

To illustrate the power of stream processing, let's consider a couple of practical scenarios.

Real-Time Analytics:

Suppose you're running a popular e-commerce website. Your site gets thousands of visitors every minute, and every interaction with your site generates data. You can use stream processing to analyze this data in real-time and gain immediate insights. For instance, you can track the number of active users on your site, the most viewed products, or the conversion rate for different product categories. You might use Kafka to ingest the data stream and Flink to calculate these metrics in real-time.

Anomaly Detection:

Imagine you're monitoring a fleet of IoT devices. Each device sends a stream of data about its status, including temperature, battery level, and error messages. You can use stream processing to detect anomalies in this data in real-time. For example, you might want to send an alert if the temperature of a device exceeds a certain threshold or if the battery level drops too quickly. In this case, you might use Kafka to ingest the data stream and Storm to implement the anomaly detection logic.

Event-Driven Applications:

You're running an online gaming platform where players can compete against each other in real-time. Every action a player takes is a potential event that could trigger other actions. For instance, when a player scores a goal, this could trigger an event that updates the scoreboard and sends a notification to all other players. Here, you could use Kafka to ingest and distribute the event data, and Samza to handle the event processing logic.

In conclusion, stream processing is a powerful tool for handling real-time data. It allows you to gain immediate insights from your data and react quickly to changes. As we generate more and more data every second, the importance of being able to process and analyze data in real-time continues to grow. Understanding the fundamental concepts of stream processing and the available technologies is crucial for working with big data.

One important consideration is the management of the state in stream processing applications. The state represents any information that is maintained across individual data records in a stream. For instance, if you're calculating a running average, the state would be the sum of the values so far and the number of values. Managing the state can be challenging, especially in the presence of failures or reprocessing needs, and is an area where stream processing frameworks provide significant help.

Moreover, time handling is a crucial aspect of stream processing. Often, you'll need to deal with the concept of 'event time' (when an event actually occurred) versus 'processing time' (when the event is observed in the system). Many stream processing tasks require grouping events by time (windowing), which can be complex due to delays or reordering in the data stream.

Also, it's essential to recognize that stream processing involves dealing with unbounded data sets. Unlike batch processing, where the data set has a fixed size, a data stream can theoretically go on forever. This presents unique challenges for resource management and scalability, and it's where the capabilities of stream processing frameworks really shine.

The beauty of stream processing lies in its ability to provide valuable insights and immediate response to the ever-flowing river of data. As you dive deeper into big data technologies, the knowledge and understanding of stream processing will open the doors to designing efficient, real-time data-intensive applications. Whether you're looking to provide personalized experiences to users, react promptly to critical situations, or drive real-time decision-making, stream processing should be an essential tool in your big data toolbox.

Module 2 Exam Questions

1. What is Big Data?

a) Data that is really important

b) Data that is too large to process using traditional methods

c) Data that is stored on a big server

d) Data that has a large number of variables

2. What is the primary function of Hadoop's HDFS component?

a) Data processing

b) Data storage

c) Data cleaning

d) Data visualization

3. Spark can process data ____

a) Only in batch mode

b) Only in real-time

c) In both batch and real-time

d) In neither batch nor real-time

4. Which of the following is NOT a NoSQL database?

a) MongoDB

b) CouchDB

c) Oracle Database

d) Cassandra

5. A data warehouse is primarily designed for ______.

a) Storing raw, unstructured data

b) Storing structured data for analysis

c) Real-time data processing

d) Data cleaning and preparation

6. In a data lake, data is stored in what form?

a) Raw, unprocessed

b) Structured, processed

c) Aggregated form

d) None of the above

7. Apache Kafka is mainly used for ____

a) Data storage

b) Batch processing

c) Stream processing

d) Data cleaning

8. Stream processing is typically used for which of the following scenarios?

a) Real-time analytics

b) Anomaly detection

c) Event-driven applications

d) All of the above

9. Which big data technology would you likely use for processing a continuous flow of data in real-time?

a) Hadoop MapReduce

b) Apache Hive

c) Apache Kafka

d) HDFS

10. Which of the following is a main component of a stream processing system?

a) Data stream

b) Stream processing engine

c) Data warehouse

d) Both a) and b)

11. Which of the following is NOT a function of the Hadoop ecosystem?

a) Real-time processing

b) Batch processing

c) Distributed storage

d) Data governance

12. What is the main difference between batch processing and real-time processing?

a) The type of data they process

b) The size of data they can handle

c) The time when they process data

d) The software they require

13. What is a key characteristic of NoSQL databases?

a) They are only suitable for small datasets

b) They require a fixed schema before data insertion

c) They are designed to handle structured data only

d) They can handle unstructured and semi-structured data

14. What does a data lakehouse seek to combine?

a) The features of data lakes and data warehouses

b) The features of NoSQL databases and data warehouses

c) The features of Hadoop and Spark

d) The features of batch and real-time processing

15. What is Apache Flink used for?

a) Data storage

b) Batch processing

c) Stream processing

d) Both b) and c)

16. Data in a data warehouse is typically _______.

a) Raw and unstructured

b) Cleaned, structured, and ready for analysis

c) In the form of streaming data

d) Stored temporarily

17. In stream processing, what is 'event time'?

a) The time when the event is processed by the system

b) The time when the event actually occurred

c) The time when the event is stored in the database

d) The time when the event is deleted from the system

18. What does the term 'state' in stream processing refer to?

a) The location of the processing server

b) The status of the processing job

c) The information that is maintained across individual data records in a stream

d) The error state of the processing system

19. Which database would you most likely use for storing JSON data?

a) MySQL

b) PostgreSQL

c) MongoDB

d) Oracle Database

20. What is one of the primary advantages of Apache Spark over Hadoop's MapReduce?

a) Spark can only process data in batch mode

b) Spark processes data slower than MapReduce

c) Spark is capable of processing data in real-time

d) Spark cannot process large datasets

Answers:

1. b
2. b
3. c
4. c
5. b
6. a
7. c
8. d
9. c
10. d
11. a
12. c
13. d
14. a
15. d
16. b
17. b
18. c
19. c
20. c

Module 3. Big Data Storage Solutions

Introduction to Big Data Storage Solutions

Welcome to Module 3 on Big Data Storage Solutions. In this module, you will learn about the essential role that storage solutions play in the Big Data ecosystem.

Big Data is characterized by the three Vs: Volume, Variety, and Velocity. With the constant growth in data volume, the diversity of data types, and the speed at which this data is generated and processed, storage becomes a crucial aspect of any Big Data initiative. Not all storage solutions are created equal, and the right choice can significantly impact the efficiency and effectiveness of your Big Data operations.

In the era of Big Data, storage is no longer a simple matter of finding a place to keep your data until it's needed. Instead, it's about choosing a solution that allows data to be efficiently stored, accessed, and processed, meeting the diverse needs of various applications and use cases. It's also about ensuring data durability and integrity, even in the face of hardware failures or network issues.

Take, for instance, a global e-commerce company. They might handle millions of transactions every day, each generating its own set of data – customer information, product details, transaction time, and more. This data is not only voluminous but also diverse, ranging from structured transaction records to semi-structured product details to unstructured customer reviews. The company needs to store all this data in a way that allows for efficient processing – whether that's for generating real-time recommendations, analyzing long-term sales trends, or identifying fraudulent transactions. This scenario illustrates the importance of selecting the appropriate Big Data storage solution.

In this module, we will delve into different storage solutions designed to handle Big Data's unique challenges. We will explore how technologies like distributed file systems, object storage, and cloud storage can help manage vast amounts of data. We'll look at the role of data compression techniques in optimizing storage usage.

By the end of this module, you will have a clear understanding of the principles underlying Big Data storage solutions and how to select the right storage technologies based on your specific requirements.

Stay tuned as we embark on this exciting journey into the world of Big Data storage solutions.

Distributed File Systems

A Distributed File System (DFS) is a type of file system that allows you to access and manage files spread across multiple machines, just as you would manage files on a single computer. The advantage of a DFS is that it allows you to store and process data across several nodes, which can be located in different geographical areas, effectively treating them as a single, unified system.

Imagine your computer's file system. You have directories, subdirectories, and files. You access them easily, and they are physically located on your computer's hard drive. Now imagine that your hard drive is not just one computer but hundreds or thousands of them, spread across different locations, maybe even different continents. That's what a distributed file system does. It makes many computers' storage look like it's just one.

Why is this so important for Big Data? Remember the three Vs of Big Data: Volume, Velocity, and Variety. Your data is not only big but also comes in quickly and in many forms. A single machine can't handle this. You need to distribute your data across many machines to manage and process it effectively.

Let's make it more practical. Suppose you have a large text file that you want to analyze, maybe it's the complete works of Shakespeare. It's too large to process on a single machine. In a distributed file system, you could split this text file into smaller parts and distribute them across multiple machines. Each machine works on its piece of the file, and the results are combined at the end. You've just sped up your analysis by a factor equal to the number of machines you used, minus some overhead for coordination and combination.

One of the most popular distributed file systems is the Hadoop Distributed File System (HDFS). HDFS is designed to handle large datasets reliably and to stream those datasets at high bandwidth to user applications. It divides large files into blocks (default size is

64MB or 128MB), distributes them across the nodes in the cluster, and also maintains replicas to ensure data reliability.

Let's consider how you might use HDFS in practice. Suppose you want to set up a Hadoop cluster on your local network for processing large datasets. You would install the Hadoop software on each machine, and one of them would be designated as the "master" node, with the others as "worker" nodes. When you want to store a file in HDFS, the system would automatically split it into blocks, distribute them among the worker nodes, and keep track of where each block is stored. If a node fails, HDFS can recover the data from a replica.

Another example of a distributed file system is the Google File System (GFS), developed by Google to support its own services. GFS also splits files into chunks and distributes them across multiple nodes, but it has unique features to support Google's particular needs.

Understanding distributed file systems is crucial for working with Big Data because they provide the foundation for many Big Data processing technologies. Whether you're using Hadoop for batch processing, Spark for in-memory processing, or other tools, chances are you'll be dealing with data stored in a distributed file system. In the following sections, we will explore other types of storage solutions and their appropriate use cases. But for now, it's important to understand that distributed file systems are a key piece of the Big Data puzzle.

Object Storage

Object storage is a strategic approach to storing data that addresses the shortcomings of other storage architectures like file and block storage. It's particularly well-suited for managing large amounts of unstructured data, making it a go-to choice for big data and cloud storage applications.

Unlike file systems that organize data into a hierarchy of files and folders, or block storage that breaks data into evenly sized blocks, object storage manages data as objects. Each object in this system includes the data, associated metadata, and a unique identifier.

The data could be anything—documents, multimedia files, applications, virtual machine images, or pieces of a larger file. The metadata is information about the data, which can include things like the creation date, content type, access control policies, and any other data you choose to include. The unique identifier allows systems and users to retrieve the object without needing to know the physical location of the data.

Consider a simple analogy to understand how object storage works. Imagine you've written several notes and you want to store them in a way that you can retrieve them later. In a traditional file system (like a filing cabinet), you'd likely organize these notes based on some sort of hierarchy or category—maybe by date, topic, or event. To find a note later, you'd have to remember the category you filed it under.

Now, imagine instead that you put each note in a box (the object), along with a list of details about the note (the metadata). You give each box a unique, random name (the identifier). To find a note later, you don't have to remember where you put it. You just ask for it by its unique name, or you can search the details in each box. That's how object storage works.

In the world of Big Data, this approach has some significant advantages. Since objects are stored in a flat address space, there's no practical limit to how many objects can be stored, making it highly scalable. The rich metadata allows for better data management and analytics. Objects can be stored redundantly across different geographic locations to improve data durability and availability.

Amazon S3 (Simple Storage Service) is a prominent example of object storage. When you store data in S3, each piece of data is stored as an object with associated metadata and a unique identifier. This enables you to retrieve the data efficiently and allows S3 to manage the underlying infrastructure on your behalf.

Let's think about how you might use Amazon S3 in a practical scenario. Suppose you're developing a photo-sharing application. You could use S3 to store the photos that your users upload. When a user uploads a photo, it's stored as an object in S3. The object's metadata could include the upload date, the user who uploaded it, and any tags or comments associated with the photo. When a user wants to view a photo, your application uses the object's identifier to retrieve it from S3. Even if your application becomes wildly popular and users upload billions of photos, S3's object storage architecture can handle the scale.

Another example is OpenStack Swift, an open-source object storage system. Like S3, Swift stores data as objects with associated metadata and identifiers. If you're operating your own data centers and want to implement object storage, you could use Swift as your object storage system.

In conclusion, object storage is a flexible, scalable, and efficient approach to managing large amounts of data. It's an important concept to understand in the context of Big Data, as it underpins many cloud storage services and big data storage solutions. As you delve deeper into Big Data concepts and technologies, you'll likely encounter object storage time and time again. So, it's crucial to understand what it is and why it's so useful.

Object storage is designed with the modern data landscape in mind, especially when dealing with massive, unstructured datasets. It offers seamless scalability, meaning you don't need to worry about running out of space or managing complex hierarchical file systems. You can keep adding data, and the system will scale automatically.

As we move further into the age of Big Data, with the Internet of Things (IoT) and other technologies generating data at an unprecedented rate, the importance of efficient and scalable data storage becomes increasingly apparent. Object storage serves this need effectively, making it an essential part of your Big Data toolkit.

Consider another practical example, a typical scenario where you're running a large website with user-generated content. This content could be anything from images and videos to blog posts and comments. With the amount of data being generated, you quickly realize that traditional file storage methods are insufficient. This is where object storage comes into play.

By using an object storage service, like Amazon S3, you can store each piece of content as an object in a flat, non-hierarchical structure, ensuring that the retrieval of these objects is efficient, regardless of the volume. Each piece of content, or 'object', can be distributed across several storage nodes, ensuring redundancy and availability.

Now, let's say you want to analyze this data for insights, such as popular content or active users. The metadata associated with each object can be leveraged to filter and analyze objects based on specific parameters, like date created, content type, or user ID. This kind of analysis would be far more difficult with traditional file or block storage.

In the case of a data-intensive application like this, object storage can significantly simplify the storage and management of large amounts of unstructured data. It provides a scalable, reliable, and cost-effective solution that can grow with your data needs.

In conclusion, object storage is an invaluable technology for handling the challenges posed by Big Data. Its flat structure, scalability, and extensive metadata capabilities make it an attractive solution for storing and managing large amounts of unstructured data. As you continue to navigate the world of Big Data, an understanding of object storage and its practical applications will serve you well.

Cloud Storage

Cloud storage is a critical component of the modern data ecosystem. It offers an alternative to on-premises data storage by enabling you to store your data on servers located in remote data centers, which are maintained by a third-party cloud provider.

With cloud storage, your data is virtually stored "in the cloud", meaning it is accessible from any location and any device with an internet connection. This not only offers convenience and flexibility but also provides a host of other benefits that can transform the way you manage your data.

One of the key advantages of cloud storage is its scalability. Unlike traditional storage systems where you need to plan and provision storage hardware in advance, with cloud storage, you can adjust your storage capacity as your needs change. If you need more storage, you can simply purchase more from your cloud provider. If your storage needs decrease, you can reduce your storage plan and only pay for what you use.

Cloud storage is also highly reliable and durable. Cloud providers often replicate your data across multiple servers in different geographic locations. This means that even if one server or data center fails, your data can still be accessed from another location.

Moreover, cloud storage providers typically offer robust security measures to protect your data, including encryption, access controls, and secure transmission protocols. This ensures your data is secure both in transit and at rest, reducing the risk of data breaches and unauthorized access.

There are several types of cloud storage, including:

- **Object storage:** As discussed in the previous section, object storage treats each piece of data as an object, storing it with associated metadata and a unique identifier. This is highly scalable and ideal for unstructured or semi-structured data. Examples include Amazon S3 and Google Cloud Storage.

- **Block storage:** This type treats data as blocks and is akin to traditional disk drives. It's excellent for applications like databases or file systems that need to perform operations on raw storage. Amazon Elastic Block Store (EBS) and Google Persistent Disk are examples.
- **File storage:** This provides a hierarchical structure, similar to the directory structure on a typical computer system, and is best suited for applications that require file system semantics. Examples include Amazon Elastic File System (EFS) and Google Cloud Filestore.

Let's consider a practical example of how cloud storage might be used. Imagine you're running a software startup, developing a mobile application that allows users to take and share photos. Initially, your user base is small, and you can manage your data storage needs on your own servers. However, as your user base grows and more and more photos are uploaded to your app, your storage needs start to exceed your capacity.

Instead of spending a significant amount of time and money to set up and manage additional servers, you can leverage a cloud storage service like Amazon S3. Your users' photos (data) can be stored as objects in the cloud, each with its own unique identifier and associated metadata, such as the user who uploaded it, the time it was uploaded, and any tags related to the photo.

With this setup, your storage capacity can easily scale as your user base grows. You no longer have to worry about running out of storage space or managing physical storage infrastructure, as these concerns are handled by the cloud provider. This allows you to focus on improving your application and providing value to your users.

Another advantage of using cloud storage in this scenario is that your data is automatically backed up and replicated across multiple locations, providing a high level of data durability and availability. Even in the event of a server failure or a natural disaster at one data center, your users' photos would not be lost and would remain accessible.

In conclusion, cloud storage is a vital tool in the Big Data landscape. Its scalability, reliability, and security features make it an ideal choice for storing and managing large amounts of data. Whether you're working with small datasets or dealing with petabytes of information, cloud storage can provide a solution that adjusts to your needs.

Imagine another scenario where you're conducting scientific research, generating huge datasets that require substantial storage space. Data could range from experimental results, images from high-resolution microscopes, to complex simulations. Storing this data on local servers could be problematic, not only due to storage limitations but also the difficulty in sharing the data with colleagues or the wider research community.

By leveraging cloud storage, you can easily store, share, and even collaborate on your data. For instance, services like Google Cloud Storage provide a secure and scalable space for your data. They also offer data sharing capabilities, allowing you to share your data with specific users, or even make it publicly available for other researchers around the world. This way, you can focus on your research while the cloud storage service takes care of data management.

Cloud storage also comes with built-in data analytics tools that you can use to glean insights from your stored data. For example, if you're storing log data from a web application on Amazon S3, you can use Amazon Athena, a serverless query service, to analyze this data. You can run SQL queries directly on your data stored in S3, without having to load it into a separate analytics system. This can save you a significant amount of time and resources in data preparation and analysis.

Furthermore, cloud storage can be an excellent solution for disaster recovery. Traditional disaster recovery solutions often involve maintaining a duplicate set of all your data and servers, which can be expensive and complicated to manage. Cloud storage allows you to back up your data to a remote location, providing a cost-effective and easy-to-manage solution for disaster recovery. In the event of a disaster, you can quickly restore your data and applications from the cloud, minimizing downtime and data loss.

In conclusion, cloud storage offers a flexible, scalable, and cost-effective solution for managing your data. It can handle vast amounts of data, provides robust security measures, and allows you to access and analyze your data from anywhere. As you navigate the world of Big Data, understanding cloud storage and how to leverage its capabilities will be an invaluable asset.

Data Compression Techniques

As you delve deeper into the world of Big Data, one of the challenges you'll often face is how to efficiently store and transmit large volumes of data. This is where data compression techniques come in handy. Data compression is a process that reduces the size of data, allowing it to be stored or transmitted more efficiently. It's a critical aspect of data storage solutions, especially in the context of Big Data.

There are two primary types of data compression: lossless and lossy.

Lossless Compression

Lossless compression is a method of data compression in which the original data can be perfectly reconstructed from the compressed data. This is the technique you'll want to use when it's important that your data remains exactly as it was before compression, such as when compressing databases, text files, or certain types of images.

For example, you might use a lossless compression algorithm like Huffman coding or Lempel-Ziv-Welch (LZW) compression. Huffman coding is often used in zip files and JPEG image compression, while LZW is used in GIF images and TIFF files.

Let's consider a practical example. Suppose you're working with a large text file containing logs from your web server. The file is so large that it's difficult to store and analyze efficiently. By using a lossless compression algorithm, you can reduce the size of the log file, making it easier to store and transfer, without losing any information. When you need to analyze the logs, you can decompress the file to its original state.

Lossy Compression

Lossy compression, on the other hand, reduces data size by eliminating less important information. This means that the decompressed data is not exactly the same as the original data, but it's often close enough to be useful. Lossy compression is often used with audio, video, and image files, where a perfect reproduction isn't necessary.

For example, the MP3 format for audio files and the JPEG format for images both use lossy compression. They achieve much greater compression than lossless methods, but at the expense of some loss of quality. However, the loss is generally small enough that it's not noticeable to human senses.

Now, let's consider a practical example with lossy compression. Suppose you're developing a mobile application that allows users to upload and share photos. Storing and transmitting high-resolution photos can be resource-intensive and slow, leading to a poor user experience. By using a lossy compression algorithm, such as the one used in the JPEG format, you can significantly reduce the size of the photos. Although some data is lost during compression, the difference is often imperceptible to the human eye, but the reduction in file size can significantly improve the speed and efficiency of your application.

Another important aspect of data compression is the concept of compression ratio, which is the ratio of the size of the compressed data to the size of the original data. The higher the compression ratio, the more the data has been reduced in size. However, a higher compression ratio can often lead to longer compression times or a loss in data quality, so it's important to find a balance that suits your specific needs.

In conclusion, data compression techniques are invaluable tools in managing and working with Big Data. They can help you make more efficient use of your storage resources, reduce data transmission times, and even improve the performance of your applications. As you continue your journey into the world of Big Data, understanding and effectively utilizing data compression techniques will be key to your success.

Module 3 Exam Questions

1. What is Big Data?

a) A large amount of structured data

b) A large amount of unstructured data

c) A large volume of data, structured or unstructured

d) A small amount of rapidly changing data

2. In the context of Big Data, what does the term 'Variety' refer to?

a) Different types of data

b) The speed at which data is generated

c) The volume of data

d) The reliability of data

3. What is a Distributed File System?

a) A file system that allows files to be stored in multiple locations

b) A file system that is stored on a single server

c) A file system that can only store text files

d) A file system that doesn't allow files to be shared

4. Hadoop Distributed File System (HDFS) is an example of:

a) Object Storage

b) Cloud Storage

c) Distributed File System

d) Data Compression Technique

5. What is Object Storage?

a) A storage architecture that manages data as objects

b) A storage architecture that manages data as blocks

c) A storage architecture that manages data as files

d) None of the above

6. Amazon S3 is an example of:

a) Object Storage

b) Distributed File System

c) Data Compression Technique

d) All of the above

7. What does Cloud Storage offer?

a) Scalability

b) Accessibility from any location

c) Cost savings

d) All of the above

8. Google Drive is an example of:

a) Cloud Storage

b) Distributed File System

c) Object Storage

d) None of the above

9. What is Data Compression?

a) A process that increases the size of data

b) A process that reduces the size of data

c) A process that protects data from being accessed without authorization

d) A process that validates the integrity of data

10. Which of the following is a type of data compression?

a) Lossless Compression

b) Lossy Compression

c) Both a and b

d) None of the above

11. JPEG uses which type of data compression?

a) Lossless Compression

b) Lossy Compression

c) Neither lossless nor lossy compression

d) Both lossless and lossy compression

12. Which type of data compression should be used when it's important for the original data to be perfectly reconstructed from the compressed data?

a) Lossless Compression

b) Lossy Compression

c) Both lossless and lossy compression

d) None of the above

13. Huffman coding is a type of:

a) Lossless Compression

b) Lossy Compression

c) Object Storage

d) Cloud Storage

14. Compression Ratio is:

a) The ratio of the size of the compressed data to the size of the original data

b) The ratio of the size of the original data to the size of the compressed data

c) The ratio of the size of the data before and after encryption

d) None of the above

15. Which one is NOT a benefit of data compression?

a) More efficient use of storage resources

b) Reduced data transmission times

c) Increased data loss

d) Improved performance of applications

16. Zip files use which type of data compression?

a) Lossless Compression

b) Lossy Compression

c) Both a and b

d) None of the above

17. TOAST in PostgreSQL is used for:

a) Data Encryption

b) Data Compression

c) Data Replication

d) Data Segregation

18. Which of the following is a practical use case for data compression?

a) Fast web page loading

b) Efficient storage of medical images

c) Improving video streaming quality

d) All of the above

19. Video streaming platforms like Netflix use what type of data compression?

a) Lossless Compression

b) Lossy Compression

c) Both a and b

d) None of the above

20. Which of the following statements about data compression is true?

a) A higher compression ratio always results in better data quality

b) Lossy compression is always preferable over lossless compression

c) Lossless compression techniques ensure that original data can be perfectly reconstructed from the compressed data

d) Data compression techniques increase the size of data

Answers:

1. c
2. a
3. a
4. c
5. a
6. a
7. d
8. a
9. b
10. c
11. b
12. a
13. a
14. a
15. c
16. a
17. b
18. d
19. b
20. c

Module 4. Big Data Processing and Analytics

Introduction to Big Data Processing and Analytics

Welcome to Module 4 on Big Data Processing and Analytics. In this module, you will learn the fundamental principles of processing and analyzing massive datasets. We will explore the various techniques, tools, and technologies that are used to transform raw data into insightful information.

Processing and analyzing big data is no small feat. The sheer volume, velocity, and variety of data that organizations generate and collect today can be overwhelming. In this context, the importance of effective big data processing and analytics cannot be overstated. It is the process of examining, cleaning, transforming, and modeling data to uncover valuable insights, identify patterns, and inform decision-making.

Imagine you're a major online retailer with millions of transactions occurring on your platform every day. Each transaction creates a data point. Each customer interaction on your website - clicks, searches, purchases, even time spent hovering over a product - generates more data. Taken individually, these pieces of data may not tell you much. But when you aggregate this data and analyze it as a whole, patterns start to emerge. You can identify which products are most popular, what times of day see the most activity, or how changes to your website's design affect user behavior.

Or consider a healthcare setting. Hospitals collect massive amounts of data from patient records, lab results, and medical imaging. By analyzing this data, healthcare providers can predict disease outbreaks, identify the effectiveness of different treatments, and personalize care for individual patients.

But how can we make sense of such vast amounts of data? This is where big data processing and analytics come in. The field has developed a range of techniques and methodologies to handle these challenges, from batch processing of data to real-time analytics, from machine learning algorithms to natural language processing, and more.

However, the process is not just about the technology. It's about asking the right questions, understanding the data you have, and knowing what to do with the information you uncover. In this module, we will not only delve into the technical aspects, but also discuss the strategic thinking and analytical mindset required to truly leverage big data.

Join us on this journey as we delve into the heart of big data, exploring the tools and techniques that can help you unearth the goldmine of insights that your data holds. By the end of this module, you will have a solid understanding of the core principles of big data processing and analytics and be equipped with the knowledge to apply these principles in real-world contexts.

Welcome to the fascinating world of Big Data Processing and Analytics. Let's begin our journey!

Batch Processing

In your journey through the world of big data, you'll encounter various processing methods. One such foundational technique is batch processing. To fully grasp its potential, let's dive deep into its definition, the mechanics behind it, and real-world applications, all with an eye towards how you might incorporate it into your own big data strategies.

Batch processing, at its core, is a way of running large volumes of data operations in a group — or "batch" — without manual intervention. This method is best suited for tasks where there is no immediate need for output, and it's more important to process large volumes of data efficiently rather than instantaneously.

Consider this example. You're running an e-commerce business, and at the end of each day, you have a plethora of data from various sources — customer transactions, website activity, inventory changes, and more. You need to process this data to gain insights, but there's no pressing need for real-time updates. Instead, your priority is to process all this data in the most resource-effective way. Here, batch processing comes into play. You can schedule the processing task to run during off-peak hours, perhaps at night when the load on your system is low.

The benefits of batch processing are multifold. First, it allows for efficient use of system resources by running during off-peak times, which can translate into cost savings. Second, it can handle vast amounts of data, making it suitable for big data contexts where data accumulates quickly. Third, since it requires no manual intervention after the initial setup, it can save your time and effort, freeing you to focus on other important tasks.

However, the practical implementation of batch processing in a big data environment requires thoughtful planning. You'll need to design and orchestrate your batch jobs carefully to prevent bottlenecks and ensure that data is processed in the correct order.

You will also need to monitor your batch processing jobs to identify any failures or performance issues.

In terms of tools, Hadoop, a widely used big data framework, has robust batch processing capabilities. Hadoop's MapReduce component, for instance, allows you to process large datasets across a cluster of computers using a simple programming model. With Hadoop, you can write your application to process the data, and the framework takes care of distributing the data and computation tasks across your cluster, handling failures, and aggregating the results.

To put this into perspective, let's consider a practical use case. Suppose you want to analyze your e-commerce website's traffic data to understand your users' behavior. You have massive log files containing data about each user's session — pages visited, time spent on each page, actions taken, and more. Using Hadoop, you can write a MapReduce job where the Map function extracts relevant information from each log entry, and the Reduce function aggregates this data to provide insights, such as the most visited pages or the average session length. You can then use these insights to improve your website and enhance your users' experience.

In conclusion, batch processing is a powerful method for handling large volumes of data in a resource-efficient way. By understanding its mechanics and applying it thoughtfully, you can unlock valuable insights from your big data and drive your business decisions. Remember, the key is to align your processing strategy with your specific needs and constraints, and batch processing is just one of the tools at your disposal.

Real-time Processing

In contrast to batch processing, another critical technique in big data processing is real-time processing. As you work with big data, there will be times when immediate insights and rapid decision-making are crucial. This is where real-time processing shines, allowing you to analyze and act on data as it arrives. Let's explore the concept, the mechanics behind it, and some practical examples to understand how you can leverage real-time processing in your big data strategies.

Real-time processing, as the name suggests, involves processing data as it is generated or received, without waiting to accumulate a large batch of data. It provides near-instantaneous insights and enables you to react quickly to events or changes in the data. This processing method is particularly useful when you need to make time-sensitive decisions, monitor systems continuously, or respond to events in near real-time.

One of the key challenges in real-time processing is managing the constant flow of incoming data. You'll need to design your system to handle the data stream efficiently, ensuring that it can process new information without getting bogged down or overwhelmed. This may involve selecting the right hardware, optimizing your algorithms, and employing techniques like data partitioning, parallel processing, and load balancing.

There are several tools and technologies available to help you implement real-time processing in a big data context. One such popular technology is Apache Kafka, a distributed streaming platform that can handle high-volume, high-throughput data streams. Kafka allows you to collect, store, and process data streams in real-time, enabling you to build responsive, data-driven applications.

To better understand real-time processing, let's consider some practical examples:

- **Fraud Detection:** In the financial industry, detecting and preventing fraud is a critical concern. Real-time processing can help you analyze transactions as they occur, identify potentially fraudulent activities, and take immediate action, such as blocking the transaction or notifying the account holder. By doing so, you can minimize the risk of financial loss and protect your customers' accounts.

- **Social Media Sentiment Analysis:** If you're running a marketing campaign or launching a new product, you might want to monitor social media platforms for real-time feedback. You can use real-time processing to analyze social media posts, comments, and messages as they are published, determining the overall sentiment towards your campaign or product. This information allows you to make quick adjustments, address concerns, and capitalize on positive feedback.

- **Internet of Things (IoT) Monitoring:** With IoT devices becoming increasingly prevalent, real-time processing plays a vital role in managing and analyzing the massive amounts of data generated by these devices. For instance, if you operate a smart city, real-time processing allows you to monitor and analyze data from sensors and connected devices across the city, enabling you to make real-time decisions about traffic management, public safety, and resource allocation.

- **Stock Market Analysis:** In the world of stock trading, every second counts. Real-time processing allows you to analyze stock market data as it arrives, providing up-to-the-minute insights into market trends and movements. This information can help you make informed, timely investment decisions and respond rapidly to market changes.

In conclusion, real-time processing is an essential technique for handling big data scenarios where immediate insights and rapid decision-making are vital. By understanding its mechanics and selecting the appropriate tools, you can build systems that process data streams efficiently and provide the real-time insights needed to make data-driven decisions. Keep in mind that the choice between batch and real-time processing depends on your specific use case and requirements, and it's crucial to select the right method for your big data project.

Machine Learning and AI in Big Data

In the ever-evolving landscape of big data, artificial intelligence (AI) and machine learning have emerged as transformative tools. They bring the ability to uncover deep insights, automate decisions, and even predict future outcomes. As we delve into this topic, we'll explore what these technologies are, how they work, and their practical applications. This exploration will equip you with an understanding of how AI and machine learning can enhance your big data strategies and create competitive advantages.

At its core, machine learning is a subset of AI that provides systems the ability to learn and improve from experience without being explicitly programmed. In the context of big data, machine learning algorithms can analyze large, complex datasets, identify patterns and trends, and make predictions based on these patterns.

The power of machine learning lies in its ability to process and learn from data that would be impossible for humans to handle manually. The more data your machine learning models have, the better they can learn and make accurate predictions. This makes machine learning a perfect fit for big data processing.

There are several types of machine learning algorithms, each with its strengths and suitable use cases. For instance, you may use supervised learning to predict future sales based on historical data, or unsupervised learning to segment your customer base into different groups. Reinforcement learning, on the other hand, can be used to train systems to make a series of decisions that maximize a reward, such as optimizing a delivery route.

AI, a broader concept, involves creating systems that can perform tasks that usually require human intelligence. This includes not only machine learning but also other aspects like natural language processing, computer vision, and robotics. AI can automate complex tasks, make intelligent decisions, and even interact with humans in a natural, human-like way.

Now, let's look at some practical examples of how you can leverage machine learning and AI in big data:

- **Predictive Analytics:** You can use machine learning to analyze historical data and predict future trends or outcomes. For instance, by analyzing past sales data, you can predict future sales, helping you make informed decisions about inventory management and marketing strategies.

- **Customer Segmentation and Personalization:** Machine learning algorithms can analyze your customer data and segment your customers into different groups based on their behavior, preferences, or other attributes. This can enable you to provide personalized experiences or recommendations, enhancing customer satisfaction and loyalty.

- **Natural Language Processing (NLP):** NLP, a subfield of AI, involves enabling computers to understand, interpret, and generate human language. You can use NLP to analyze customer reviews, social media posts, or other text data, gaining insights into customer sentiment, trends, and themes.

- **Anomaly Detection:** Machine learning can help you detect anomalies or outliers in your data. This can be particularly useful for identifying fraudulent transactions, network intrusions, or system failures. By detecting such anomalies in real time, you can take immediate action to mitigate any potential issues.

- **Chatbots and Virtual Assistants:** AI can power chatbots and virtual assistants that can interact with users in a natural, human-like way. These can provide 24/7 customer support, answer frequently asked questions, and even perform tasks like scheduling appointments or placing orders.

In conclusion, machine learning and AI offer powerful tools for processing and analyzing big data. They bring the potential to uncover deep insights, make predictions, automate decisions, and much more. By understanding these technologies and their applications, you can harness their potential to enhance your big data strategies and gain a competitive edge. Remember, the key is to align these technologies with your specific needs and goals, and to continually evaluate and refine your approach as you gain more data and insights.

Text Analytics and Natural Language Processing

As you delve into the world of big data, you'll often encounter unstructured data, one of the most common forms of which is text. This is where the fields of text analytics and Natural Language Processing (NLP) come into play. In this section, we'll explore these two related areas, their techniques, and their applications in the context of big data.

Text analytics refers to the process of deriving high-quality information from text. It involves structuring the input text, deriving patterns within the structured data, and finally evaluating and interpreting the output. It's used on a large scale to uncover key insights, trends, and patterns that are not immediately apparent, especially when dealing with vast amounts of text data.

NLP, on the other hand, is a subfield of artificial intelligence that focuses on the interaction between computers and humans through natural language. The ultimate objective of NLP is to read, decipher, understand, and make sense of human language in a valuable way. It involves several complex tasks such as understanding syntax (the grammatical structure), semantics (the meaning), and even context.

Now, how do these two fields apply to big data? Well, a considerable portion of the data generated every day is unstructured text data - think of social media posts, customer reviews, emails, documents, and more. Text analytics and NLP provide the tools to analyze this data, derive meaningful insights, and make data-driven decisions.

Let's look at some practical examples of how you can leverage text analytics and NLP in your big data journey:

- **Sentiment Analysis:** This involves determining the emotional tone behind words to gain an understanding of the attitudes, opinions, and emotions expressed within an online mention. For instance, by analyzing social media posts or customer reviews, you can gain insights into how your customers feel about your products, services, or brand.

- **Customer Service Automation:** You can use NLP to build chatbots or virtual assistants that can understand and respond to customer queries in a natural, human-like way. This can help you provide 24/7 customer support, handle more queries at once, and free up your human agents to deal with more complex issues.

- **Information Extraction:** You can extract useful information or insights from large collections of documents or web pages. For example, you might use text analytics to analyze legal documents and extract key terms, parties, or obligations.

- **Topic Modeling:** If you have a large collection of documents, you can use text analytics to group them into topics. This can help you understand the main themes in your data, organize your documents, and find relevant information more quickly.

- **Language Translation:** NLP can help you translate text from one language to another, making your content accessible to users in different countries. This can be particularly useful if you operate globally or have a multilingual user base.

- **Speech Recognition:** NLP can convert spoken language into written text, enabling you to transcribe audio, build voice-controlled applications, or even analyze spoken data.

In conclusion, text analytics and NLP provide powerful tools for dealing with unstructured text data. They enable you to uncover insights, automate tasks, and make data-driven decisions, enhancing your big data strategies. However, text data can be complex and messy, and it's crucial to clean and preprocess your data effectively before analysis. By understanding these techniques and their applications, you can unlock the full potential of your text data and gain a competitive edge.

Graph Analytics

As you continue your exploration into the world of big data, you'll soon realize that not all data is created equal. Different types of data require different types of processing and analytics techniques. One such type of data that has gained considerable attention in recent years is graph data, and with it, the field of graph analytics.

A graph, in the data context, is a collection of vertices (or nodes) and edges. Each node represents an entity, and each edge represents a relationship or connection between two entities. Graphs are incredibly versatile and can represent a wide variety of systems from social networks, web pages, biological networks, to logistics and more.

Graph analytics, also known as network analysis, involves extracting insights from graph data. It provides a way to understand the relationships and patterns among nodes and edges in the graph, offering you a more comprehensive view of your data.

The beauty of graph analytics is its ability to simplify the complexity of relationships in data and provide insights that might otherwise be hard to spot. This is especially useful when dealing with big data, where the volume and complexity of relationships can be overwhelming.

Now, let's take a look at some practical ways you could apply graph analytics to your big data:

- **Social Network Analysis:** Imagine you're managing a social media platform. Graph analytics can help you understand the relationships and interactions between your users. You could identify influencers in your network (those with a high degree of connections), detect communities of users with similar interests, or even predict future connections.

- **Fraud Detection:** If you're in a financial institution, graph analytics can play a significant role in detecting fraud. By representing transactions as a graph, you can identify unusual patterns or anomalies that might indicate fraudulent activity. For instance, a sudden surge in transactions between a group of accounts might suggest a fraud scheme.

- **Recommendation Systems:** If you're in e-commerce or any customer-facing business, graph analytics can enhance your recommendation systems. By understanding the relationships between customers and products (such as purchase history, browsing behavior, etc.), you can recommend products that a customer is likely to be interested in.

- **Logistics and Supply Chain Optimization:** Suppose you're managing a logistics network. In that case, graph analytics can help you optimize your routes and reduce costs. Each location can be a node, and each route an edge, and you can use graph analytics to find the shortest path, identify bottlenecks, or balance loads across your network.

- **Web Search:** If you're dealing with web data, graph analytics can improve your search algorithms. Web pages can be nodes, and links between them edges. A famous example of this is Google's PageRank algorithm, which ranks web pages based on their relationships with other pages.

As you can see, graph analytics provides a unique perspective on your data, emphasizing relationships and interactions. However, working with graph data can be challenging, as it often involves complex data structures and algorithms. It's crucial to understand these complexities and choose the right tools and techniques for your needs.

Distributed Computing

As you venture into big data processing and analytics, it's essential to grasp the concept of distributed computing. Dealing with large datasets often requires processing power beyond the capabilities of a single machine. Distributed computing offers a solution by splitting the workload across multiple machines or clusters, allowing you to handle and process big data more efficiently. In this section, we'll cover the fundamentals of distributed systems, explore various distributed computing models, and discuss the challenges associated with distributed computing.

Understanding Distributed Systems

A distributed system consists of multiple independent computers, known as nodes, that communicate and coordinate with each other to achieve a common goal. The nodes in a distributed system often have their own memory, storage, and processing capabilities, and they are connected via a network. The primary objective of a distributed system is to distribute tasks and data across multiple nodes, enabling you to tackle problems that would be too resource-intensive or time-consuming for a single machine.

In the context of big data, distributed systems are crucial for several reasons:

- **Scalability:** As the volume of data grows, distributed systems can easily scale by adding more nodes to handle the increased workload.

- **Fault Tolerance:** Distributed systems can continue to function even if individual nodes fail, ensuring that your data processing and analytics tasks can proceed without interruption.

- **Load Balancing:** By distributing tasks and data across multiple nodes, you can balance the load and ensure that no single node becomes a bottleneck.

Distributed Computing Models

There are several distributed computing models that you can choose from, depending on your specific requirements and the nature of your big data tasks. Some popular models include:

- **MapReduce:** This is a programming model popularized by Hadoop for processing and generating large datasets. It consists of two primary functions, Map and Reduce. The Map function processes input data and generates a set of key-value pairs. The Reduce function then processes these key-value pairs and generates the final output. MapReduce is highly scalable and fault-tolerant, making it well-suited for big data processing.

- **Data Parallelism:** In this model, the same operation is applied to different subsets of data simultaneously. This is particularly useful when dealing with large datasets that can be partitioned and processed in parallel. For example, if you're training a machine learning model on a large dataset, you can divide the dataset into smaller chunks and process each chunk on a different node.

- **Task Parallelism:** In task parallelism, different tasks are executed concurrently on separate nodes. This approach is well-suited for workflows where tasks can be performed independently of each other. For instance, if you have multiple independent data transformation tasks to perform on your dataset, you can distribute them across different nodes.

- **Stream Processing:** This model is designed for processing continuous data streams in real-time. Data is ingested, processed, and analyzed as it arrives, allowing you to extract insights and make decisions quickly. Apache Kafka and Apache Flink are popular tools for stream processing.

Challenges in Distributed Computing

While distributed computing offers numerous advantages for big data processing, it also comes with its own set of challenges:

- **Data Consistency:** Ensuring that data remains consistent across all nodes in a distributed system can be a complex task. This is particularly true when multiple nodes are updating data simultaneously. You may need to implement synchronization and consistency mechanisms to ensure data integrity.

- **Latency:** As tasks and data are distributed across multiple nodes, communication between nodes becomes a critical factor. Network latency can impact the overall performance of your distributed system, so it's essential to optimize your network topology and minimize communication overhead.

- **Fault Tolerance:** Dealing with node failures is a critical aspect of distributed computing. You'll need to implement strategies for detecting failures and recovering from them, such as replication and checkpointing.

- **Complexity:** Designing, implementing, and managing distributed systems can be complex. You'll need to handle issues like task scheduling, load balancing, and data partitioning, among others. Thankfully, there are several distributed computing frameworks like Hadoop and Spark that can handle much of this complexity for you.

As you delve deeper into distributed computing, keep these concepts and challenges in mind. Understanding the intricacies of distributed systems will enable you to leverage them effectively for your big data processing and analytics tasks, allowing you to extract valuable insights from your data efficiently and reliably.

Data Cleaning and Preprocessing

Navigating the world of big data isn't just about having vast amounts of information at your disposal. It's about making sure that your data is clean, reliable, and ready for analysis. Data cleaning and preprocessing is an essential step in any big data pipeline, ensuring the quality and accuracy of your data before it's subjected to further processing or analysis. In this section, we'll look at the importance of data cleaning, common data quality issues you may encounter, and various techniques you can use to clean and preprocess your data.

Importance of Data Cleaning

Data is the foundation of any data-driven decision-making process. If your data is inaccurate, incomplete, or inconsistent, the insights and conclusions drawn from that data will likely be flawed. This is where data cleaning comes into play. Data cleaning involves detecting and correcting (or removing) errors and inconsistencies from your data to improve its quality.

Here's why data cleaning is so important:

- **Accuracy:** Clean data leads to accurate analysis. By removing errors and inconsistencies, data cleaning ensures that your data accurately reflects the real-world phenomena it's supposed to represent.

- **Confidence:** When your data is clean, you can trust the insights and predictions generated from it. This trust is essential for making informed decisions based on your data.

- **Efficiency:** Clean data can be processed and analyzed more efficiently. Errors and inconsistencies can cause your analysis algorithms to produce incorrect results or even fail.

Common Data Quality Issues

As you work with big data, you're likely to encounter various data quality issues. Some of the most common ones include:

- **Missing Values:** Not all data sources will have complete information for every record. Missing values can lead to biased or incorrect results during analysis.

- **Duplicate Entries:** Duplicates can occur for a variety of reasons, such as data entry errors or merging data from multiple sources. They can skew your analysis and lead to incorrect conclusions.

- **Inconsistent Data:** This could include inconsistencies in data formatting, naming conventions, or measurement units, making it difficult to aggregate or compare data.

- **Outliers:** Outliers are data points that significantly differ from the rest of your data. While some outliers represent valuable insights, others are just noise resulting from errors or anomalies.

Techniques for Data Cleaning and Preprocessing

There are several techniques you can use to clean and preprocess your data:

- **Handling Missing Values:** Depending on the nature of your data and the reason for the missing values, you can choose to fill in the missing values with a default value or a calculated value (like the mean, median, or mode), or you can choose to remove the records with missing values entirely.

- **Removing Duplicates:** This involves identifying and removing duplicate records from your data. Most data processing tools provide functionality for detecting and eliminating duplicates.

- **Data Formatting and Transformation:** This involves standardizing and transforming your data to a consistent format. This could include tasks like converting data types, standardizing date formats, or mapping categorical values to numerical values.
- **Outlier Detection and Treatment:** Outliers can be detected using statistical methods like the z-score or IQR method. Once detected, you can choose to remove outliers or treat them based on their nature and impact on your analysis.
- **Feature Scaling/Normalization:** When your dataset features have different scales, you might want to scale them to a uniform range, especially for certain machine learning algorithms. Methods include Min-Max Scaling, Standardization (Z-score normalization), and Robust Scaling.

Remember, data cleaning and preprocessing isn't a one-size-fits-all process. The techniques you choose to use will depend on your data, your use case, and the specific requirements and constraints of your analytics project.

Advanced Analytics Techniques

With the basics of big data processing and analytics under your belt, it's time to delve into some of the more advanced analytics techniques. These techniques, which include time-series analysis, complex event processing, and spatial data analysis, can provide you with deeper insights and more nuanced understanding of your data. In this section, we'll explore each of these techniques in detail, and provide you with practical examples to help you understand their applications.

Time-series Analysis

A time series is a sequence of data points collected or recorded in time order. This type of data is prevalent in many fields, including finance (stock prices), meteorology (temperature over time), and even social media (tweets per day). Time-series analysis involves using statistical techniques to model and explain the behavior and underlying patterns in the data.

For example, if you're analyzing web server logs, you could use time-series analysis to detect patterns and trends in web traffic over time. This could help you anticipate future traffic patterns and ensure that your servers can handle the load. Or, if you're analyzing stock prices, time-series analysis could help you identify trends and make predictions about future prices.

There are various methods for time-series analysis, including:

- Autoregressive Integrated Moving Average (ARIMA)
- Seasonal Decomposition of Time Series (STL)
- Exponential Smoothing

Each of these methods has its strengths and weaknesses, and the choice of method depends on the nature of your time series data and the specific problem you're trying to solve.

Complex Event Processing

Complex event processing (CEP) is a technique used to analyze and identify meaningful patterns, sequences, and relationships among multiple events happening across various sources in real-time. This kind of analysis is particularly useful in domains where you need to make instant decisions based on the event patterns, such as fraud detection in banking, network security monitoring, or real-time stock trading.

For instance, consider a credit card fraud detection system. A single transaction might not seem fraudulent on its own. But if CEP identifies a sequence of transactions occurring in a short period across different geographic locations, it could flag this pattern as potential fraud.

CEP involves the following steps:

- **Event Collection:** Gathering the raw data or events from various sources.
- **Event Processing:** This could include filtering, aggregation, transformation, or enrichment of the raw events.
- **Pattern Detection:** Identifying the meaningful patterns or sequences among the events.
- **Action:** Based on the patterns identified, an appropriate action is taken.

Spatial Data Analysis

Spatial data, also known as geospatial data, includes any data that has a geographical or spatial aspect. This could be anything from the coordinates of a restaurant, to the boundaries of a city, to the routes of delivery trucks. Spatial data analysis involves analyzing this data to understand the geographical relationships, patterns, and trends.

For example, if you're a retailer with multiple store locations, you could use spatial data analysis to identify the areas with the highest sales. This could help you make more informed decisions about where to open new stores, or where to focus your marketing efforts.

Spatial data analysis can involve a range of techniques, including:

- **Geocoding:** Converting addresses into geographic coordinates.
- **Spatial Clustering:** Grouping spatial data based on their geographical proximity or other related attributes.
- **Hotspot Analysis:** Identifying areas of high activity or values.

Spatial data analysis often involves using specialized software or libraries, such as Geographic Information System (GIS) software, or spatial analysis libraries in Python or R.

As you explore these advanced analytics techniques, remember that the goal is to extract meaningful insights from your data. These techniques provide you with additional tools and perspectives, but the most important thing is to choose the techniques that are most suited to your data and your specific analytical goals.

Emerging Trends in Big Data Processing and Analytics

As you navigate the complex landscape of big data processing and analytics, it's important to keep an eye on emerging trends. These trends often represent the cutting-edge of technology and methodology, providing new ways to handle, analyze, and extract insights from data. In this section, we'll explore three such emerging trends: federated learning, automated machine learning (AutoML), and explainable AI (XAI).

Federated Learning

As you know, machine learning models improve with data. The more data they can learn from, the better their predictive performance. However, privacy and security concerns often limit the amount of data that can be centrally collected for training these models. Enter federated learning, a decentralized approach to machine learning.

In federated learning, a model is sent to where the data resides (like your smartphone or a local server) and learns from the data in place. The model updates (not the data) are then sent back to a central server where they're combined with updates from other models. This way, all the learning is done locally on your own device, and only non-sensitive model updates are shared, preserving privacy.

Imagine you're working on improving a voice recognition model for a personal assistant app on your smartphone. With federated learning, the voice recognition model can learn from your voice data directly on your device without sending any sensitive voice recordings to a central server.

Automated Machine Learning (AutoML)

Machine learning model development involves many steps: preprocessing data, selecting features, choosing a model, tuning hyperparameters, and validating the model. This can be a complex and time-consuming process, even for experienced data scientists. This is where automated machine learning, or AutoML, comes in.

AutoML aims to automate the process of applying machine learning, making it more accessible and efficient. With AutoML, you specify the data and the problem, and the system automatically selects the best preprocessing techniques, features, models, and hyperparameters. It also validates the model and provides you with a performance evaluation.

For instance, say you're trying to predict customer churn based on a variety of customer attributes. With AutoML, you can simply feed in your data, specify that you're trying to predict churn, and let the system handle the rest. It'll automatically build, evaluate, and select the best performing model for your data.

Explainable AI (XAI)

As AI and machine learning models become more complex, understanding how they make their decisions becomes more difficult. This lack of transparency, or "black box" problem, can be a significant issue, especially in high-stakes fields like healthcare or finance where it's essential to understand why a model made a particular prediction.

Explainable AI (XAI) aims to make AI decision-making transparent and understandable to humans. This involves developing models that provide clear explanations for their decisions, or creating methods that can interpret the decisions of complex models.

For example, you might use an XAI technique to interpret a deep learning model used for diagnosing diseases based on medical images. The XAI technique could provide an explanation of the model's diagnosis, such as highlighting the areas of the image that the model found most indicative of the disease.

These emerging trends represent exciting developments in the field of big data processing and analytics. By staying informed about these trends, you can ensure that you're equipped with the latest tools and techniques to extract the most value from your data.

Module 4 Exam Questions

1. What is Big Data Processing?

a) A method of creating large volumes of data

b) A technique for visualizing data

c) The act of taking large volumes of data and performing operations to generate insights

d) A security protocol for data

2. What does Batch Processing involve?

a) Processing data in real-time as it arrives

b) Processing data in large groups or 'batches'

c) Visualizing data

d) Encrypting data

3. Which of the following is a key characteristic of Real-time Processing?

a) Data is processed immediately as it arrives

b) Data is processed in large groups or batches

c) The processing of data is delayed

d) Data is stored for future processing

4. What is the main goal of Text Analytics?

a) To create large volumes of text data

b) To analyze and draw insights from text data

c) To visualize text data

d) To store text data

5. Graph Analytics is primarily used to analyze what type of data?

a) Textual data

b) Numerical data

c) Data in the form of graphs and networks

d) None of the above

6. What is a primary feature of Distributed Systems in the context of Distributed Computing?

a) Data is stored in one central location

b) Processing power is sourced from a single machine

c) Data and processing power are distributed across multiple machines or systems

d) None of the above

7. Why is Data Cleaning and Preprocessing important in Big Data Processing and Analytics?

a) It removes irrelevant data

b) It enhances the quality of data, making it suitable for analysis

c) It helps in making data consistent

d) All of the above

8. Time-series Analysis is used for analyzing data that:

a) Has been collected over a period of time

b) Is collected in real-time

c) Is highly structured

d) None of the above

9. What does Complex Event Processing (CEP) help you to identify?

a) Meaningful patterns and relationships among events

b) The most recent event in a sequence

c) Events that have occurred in the past

d) None of the above

10. Which of the following is a technique used in Spatial Data Analysis?

a) Geocoding

b) Autocorrelation

c) Spatial Clustering

d) All of the above

11. Federated Learning is a technique that:

a) Centralizes all data for model training

b) Sends the data to a model for training

c) Sends the model to where the data resides for training

d) None of the above

12. What is the primary goal of Automated Machine Learning (AutoML)?

a) To make machine learning more accessible and efficient

b) To decentralize machine learning

c) To increase the amount of data required for machine learning

d) None of the above

13. Explainable AI (XAI) is designed to make AI decision-making:

a) More complex

b) Transparent and understandable to humans

c) Less accessible

d) None of the above

14. In which type of processing is data processed in large groups?

a) Real-time processing

b) Batch processing

c) Federated learning

d) None of the above

15. What is a key feature of real-time processing?

a) There is a delay in data processing

b) Data is processed as it arrives

c) Data is processed in large groups

d) None of the above

16. What is the primary focus of machine learning in big data?

a) Generating large volumes of data

b) Analyzing and drawing insights from the data

c) Visualizing data

d) None of the above

17. Which of the following is a trend in Big Data Processing and Analytics?

a) Federated learning

b) Automated Machine Learning (AutoML)

c) Explainable AI (XAI)

d) All of the above

18. What is the key objective of data preprocessing?

a) To increase the volume of data

b) To enhance the quality and consistency of the data

c) To visualize the data

d) None of the above

19. Which of the following is a method used in Text Analytics?

a) Sentiment Analysis

b) Topic Modeling

c) Named Entity Recognition

d) All of the above

20. What is the primary use of Graph Analytics?

a) To analyze numerical data

b) To analyze and draw insights from data represented as graphs

c) To visualize textual data

d) None of the above

Answers:

1. c
2. b
3. a
4. b
5. c
6. c
7. d
8. a
9. a
10. d
11. c
12. a
13. b
14. b
15. b
16. b
17. d
18. b
19. d
20. b

Module 5. Big Data Visualization

Introduction to Big Data Visualization

Welcome to Module 5 on Big Data Visualization. In this module, you will learn about one of the most crucial and influential aspects of the big data landscape – visualization. Visualization not only enhances our comprehension of complex datasets but also enables us to communicate insights effectively. It is the art and science of making data comprehensible and actionable to both technical and non-technical audiences.

Data is growing exponentially in today's digital world, and making sense of this sea of information requires more than just analytical skills; it demands the ability to present this data in a visually appealing and comprehensible manner. The human brain processes visual information far more efficiently than raw, tabular data, which is why data visualization is an essential component of modern data science and big data analytics.

In this module, we will dive deep into the principles and practices of data visualization, exploring how it can transform raw data into meaningful insights. Data visualization is not merely about creating visually appealing charts and graphs. It is a multidisciplinary field that combines design, psychology, and storytelling to present data in a way that resonates with our human perception and cognition.

Consider, for example, how a city's government might use data visualization to understand and communicate traffic patterns. By visualizing the flow of traffic at different times of day, decision-makers can identify bottlenecks, plan for road improvements, and determine where to allocate resources. Similarly, in the healthcare sector, visualizing patient data can help in identifying patterns and trends in disease outbreaks, thereby enabling faster and more effective responses.

As we journey through this module, you will gain an understanding of the power and potential of data visualization, its significance in the context of big data, and the tools and techniques that can help you create impactful visual narratives from your data. By the end of this module, you will be able to harness the power of visualization to bring your big data to life, transforming it from mere numbers into compelling stories that drive decision-making and innovation.

Prepare to immerse yourself in the fascinating world of big data visualization, where data meets design, and analytics meet art. Let's embark on this journey to unlock the hidden narratives within our data and inspire data-driven decisions.

Principles of Data Visualization

In this section we will explore the fundamental principles of data visualization. As we delve into this topic, you will find that these principles are not just guidelines, but rather vital tools that can help you transform data into powerful visual stories.

Understand Your Audience and Purpose

The first and perhaps most important principle is understanding your audience and the purpose of your visualization. Are you creating a visualization for a technical team, or will a non-technical audience be consuming it? Are you trying to convey a specific point, or are you facilitating exploratory data analysis? When you understand your audience's level of data literacy and the objective of your visualization, you can choose the most effective visual forms and design elements.

Choose the Appropriate Type of Visualization

The type of visualization you select should align with the nature of your data and the story you want to tell. Bar charts, for instance, are excellent for comparing categorical data. Line graphs effectively represent trends over time. Scatter plots help in revealing the correlation between two variables. There are many types of visualizations to choose from, and understanding when to use each one is key to presenting your data effectively.

Simplicity is Key

The best visualizations are often the simplest. Avoid cluttering your visualization with unnecessary elements that could distract from the data. Each element in your visualization, from lines to colors, should serve a purpose. If an element does not enhance understanding or provide necessary context, consider removing it.

Use Colors and Sizes Wisely

Colors and sizes are powerful visual cues. Use them wisely to emphasize important points and patterns. However, be mindful of the potential pitfalls. For instance, using too many bright colors can be overwhelming and confusing. Similarly, using size to represent a variable can lead to misinterpretation if not done carefully.

Maintain Consistency

Consistency in your visualization helps your audience understand the data more easily. This can mean consistently using the same color to represent a specific category across multiple visualizations, or maintaining the same scale when comparing different datasets.

Ensure Accuracy and Integrity

As a data professional, it's your responsibility to represent the data accurately. Even small missteps can lead to large misinterpretations. Avoid cherry-picking data or manipulating the visualization to exaggerate or downplay certain aspects of the data.

Let's look at a practical example. Suppose you're analyzing a dataset about the average monthly rainfall in your city over the past year. You want to create a visualization that accurately represents this data and helps you identify any trends or patterns.

First, you must understand your audience and purpose. Let's say your audience is a group of city planners who need this information to plan for potential water shortages or flooding. They may not be data experts, so your visualization needs to be clear and straightforward.

A line graph would be an appropriate choice for visualizing this data, as it's excellent for showing changes over time. On the X-axis, you would have the months of the year, and on the Y-axis, the amount of rainfall.

For the design, you choose simplicity. You opt for a clean white background, with the line in a clear, bright color that stands out, like blue - a color often associated with water. You avoid any unnecessary elements, like gridlines or extra annotations, that could clutter the visualization.

You decide to use size and color to emphasize the month with the highest rainfall. This month's point on the graph is slightly larger and a darker shade of blue, immediately drawing attention.

To maintain consistency, you ensure that the scale on your Y-axis is consistent, with equal intervals representing the same amount of rainfall.

Finally, you check and double-check your data and your graph to ensure accuracy. You make sure that each point on the graph correctly corresponds to the amount of rainfall that month.

This is just one example of how these principles can guide you in creating effective data visualizations. As you continue to learn and experiment, you'll find that these principles are not rigid rules, but flexible guidelines that can help you make the most of your data and share its insights in the most compelling way possible.

Popular Visualization Tools

Now, we will explore popular tools for data visualization. These tools offer a wide array of functionalities to make the visualization process more efficient and effective. From creating simple charts to complex interactive dashboards, you'll find the right tools to bring your data to life.

Tableau

Tableau is a powerful data visualization tool widely used across industries. It is known for its user-friendly interface and the ability to create sophisticated visualizations and interactive dashboards. With Tableau, you can connect directly to your data sources, whether they are flat files, databases, or even big data sources. You can also perform basic data transformations within Tableau itself.

For example, if you're working with a sales dataset, you can use Tableau to create an interactive dashboard that displays sales by region, product category, and time period. This allows you to easily explore your data and uncover insights such as seasonal trends or regional differences.

Microsoft Power BI

Microsoft Power BI is another popular tool for data visualization, particularly among businesses that heavily use other Microsoft products. Power BI allows you to create a wide variety of visualizations and reports, and its integration with the Microsoft suite allows for seamless data import and export from Excel, Azure, and other Microsoft tools.

Imagine you are a project manager tracking the progress of various tasks. You could use Power BI to create a Gantt chart to visualize your project timeline. You could also create an interactive dashboard that allows you to drill down into specific tasks or team members, helping you manage your project more effectively.

Python Libraries: Matplotlib, Seaborn, Plotly

For those comfortable with programming, Python offers several powerful libraries for data visualization. Matplotlib is the most commonly used and forms the basis for other Python visualization libraries. It's great for creating basic graphs like line charts, bar charts, histograms, and scatterplots.

Seaborn is built on top of Matplotlib and integrates well with Pandas data structures. It's particularly useful for statistical visualizations and it comes with some built-in datasets that you can use to practice your skills.

Plotly, another Python library, stands out for its interactive visualizations. It supports a wide array of charts, including some more complex types like contour plots, dendrograms, and 3D charts. You can use these tools to explore your data in a hands-on, interactive way.

D3.js

If you're comfortable with JavaScript and HTML, D3.js offers a powerful and flexible way to create custom visualizations. D3 stands for Data-Driven Documents, and it allows you to bind arbitrary data to a Document Object Model (DOM), and then apply data-driven transformations to the document.

For example, if you have a dataset about global temperatures, you could use D3.js to create an interactive map that shows how temperatures have changed over time. Users could zoom in on specific regions, or use a slider to see changes year by year.

Excel

Excel, while not a specialized visualization tool, is widely used for creating basic charts and graphs, and it's a good starting point for beginners. Its graphical interface is easy to use, and it's capable of creating a variety of chart types, from column and line charts to pie and area charts. Despite its simplicity, Excel can be quite powerful when used effectively.

For instance, if you're tracking your personal expenses, you could use Excel to create a pie chart that shows your spending by category. This can give you a clear visual overview of where your money is going, helping you manage your finances more effectively.

Remember, the best tool for you depends on your specific needs and skills. Whether you prefer a drag-and-drop interface or coding your visualizations from scratch, there's a tool out there that's right for you. As you gain more experience with these tools, you'll be able to create increasingly sophisticated and impactful visualizations.

Interactive Dashboards and Reporting

Let's delve into the world of interactive dashboards and reporting. This is where data visualization gets truly dynamic, allowing you to engage directly with your data and uncover deeper insights.

Interactive Dashboards

An interactive dashboard is a data management tool that visually tracks, analyzes, and displays key performance indicators (KPIs), metrics, and key data points. Unlike static visualizations, interactive dashboards allow you to manipulate the data in real-time, enabling a more exploratory and immersive approach to data analysis.

Imagine you are running an e-commerce website. An interactive dashboard can help you monitor your site's performance in real-time. You could have a map showing sales by location, a line chart displaying daily visitors, a bar graph comparing sales across different product categories, and a gauge showing the current average customer review score.

This isn't just a collection of separate charts, though. Because it's interactive, you can filter and drill down into the data. You could select a specific region on your sales map and see how other metrics change for that region. You could click on a specific category in your bar graph to see how its sales have changed over time. You could adjust the time frame to see how these metrics have evolved.

Reporting

Reporting, in the context of data visualization, is the process of organizing data into informational summaries to monitor how different areas of a business are performing. These reports can be automatically generated and distributed on a regular schedule or created ad hoc as required.

Building on our e-commerce example, you might want to create a weekly sales report. This could include a summary of total sales, average customer review scores, website traffic, and any significant changes in these metrics. You could also include charts showing sales trends over time and a ranking of your top-selling products.

Tools for Interactive Dashboards and Reporting

There are numerous tools available for creating interactive dashboards and reports. As we discussed in the previous section, both Tableau and Power BI are powerful tools for this purpose.

With Tableau, for example, you can create dashboards that allow users to filter and drill down into the data. You can also add interactive elements like drop-down lists, sliders, and checkboxes that let users customize the view to their needs. Plus, Tableau's device designer feature lets you optimize your dashboards for various devices, ensuring a seamless user experience whether your dashboard is viewed on a desktop, tablet, or phone.

Power BI also offers a wide array of interactive features. You can create slicers that let users filter the data, or add drillthrough filters that let users click on a data point to see more detailed information. Power BI also lets you publish your dashboards and reports online, making it easy to share them with others.

Both Tableau and Power BI also support automated reporting. You can schedule your reports to be generated and distributed on a regular basis, ensuring that stakeholders always have the latest information at their fingertips.

Whether you're monitoring your business's performance, tracking a project, or analyzing a complex dataset, interactive dashboards and reporting can help you engage with your data in a more meaningful way.

Geospatial Data Visualization

In this part, we will delve into geospatial data visualization. This form of visualization involves the graphical representation of data that has a geographical or spatial component. By adding a spatial dimension to your data, you can uncover patterns and insights that might otherwise go unnoticed.

Understanding Geospatial Data

Geospatial data, sometimes known as spatial data, is data that is associated with a specific location. This could be anything from the coordinates of a city, the boundaries of a country, or the route of a migrating animal. Geospatial data is unique because the spatial component allows you to analyze data in the context of physical locations or landscapes.

Methods of Geospatial Data Visualization

There are several ways to visualize geospatial data, and the best method depends on your specific data and what you want to convey. Here are a few common methods:

- **Maps:** The most common method is simply plotting your data on a map. This could be a dot map, where each dot represents a data point, or a choropleth map, where regions are colored or patterned in relation to a data variable.

- **Heatmaps:** Heatmaps can be used to show the density of points in a particular area or to indicate the intensity of a certain variable.

- **3D Models:** For more complex data, you might create a 3D model. This could be a surface plot showing the topography of a landscape, or a 3D bar chart showing population density in different regions.

- **Geospatial Networks:** If your data involves connections between different locations, you might create a geospatial network. This could show anything from transportation routes to social connections between different locations.

Geospatial Visualization Tools

There are various tools available for geospatial data visualization. For simpler visualizations, you might use a tool like Excel or Tableau, which allow you to create maps and other basic geospatial visualizations. Python also offers several libraries for geospatial data visualization, including Geopandas, Bokeh, and Folium.

For more complex or large-scale geospatial data, you might use a more specialized tool like QGIS or ArcGIS. These tools offer advanced features for geospatial analysis and visualization, allowing you to work with large datasets and create sophisticated visualizations.

Practical Examples

Let's say you're studying the spread of a disease. You have data on the number of cases in different cities around the world. You could plot this data on a map, with each city represented by a dot. You could then size the dots based on the number of cases, creating a bubble map that clearly shows where the disease is most prevalent.

Or, suppose you're working with a dataset that includes GPS tracking data for a fleet of delivery vehicles. You could use this data to create a geospatial network showing the routes each vehicle takes. By adding a time component, you could even animate this network to show how the vehicles move throughout the day.

Geospatial data visualization is a powerful tool that can help you understand your data in new and meaningful ways. By linking your data to the physical world, you can uncover patterns and trends that might not be evident from the numbers alone.

Storytelling with Data

Welcome to Section 5.5. In this section, we will explore the art of storytelling with data. The goal of data visualization is not only to create visually appealing charts and graphs but also to convey a clear message or story that resonates with your audience. Here, we will guide you through selecting the right visualizations, using captions and annotations effectively, and structuring your presentation for maximum impact.

Choosing the Right Visualizations

The first step in storytelling with data is selecting visualizations that best support your narrative. This involves considering your data and the key points you want to convey. Here are a few tips to help you choose the right visualizations:

- **Understand your data:** Before deciding on a visualization, ensure you have a clear understanding of your data, its structure, and any patterns or trends that you want to highlight.

- **Identify your key message:** Consider the main takeaway you want your audience to grasp. Your visualization should be tailored to emphasize this message and make it clear to your audience.

- **Keep it simple:** While it can be tempting to create complex visualizations with numerous data points, it is often better to opt for simplicity. Simple visualizations are more likely to be understood quickly and effectively.

- **Experiment with different visualizations:** Try different visualization types to see which ones best convey your message. You may find that a line chart works better for showing trends, while a bar chart is more effective at comparing values.

Using Captions and Annotations Effectively

Captions and annotations can be powerful tools for enhancing your data story. They provide context, highlight important points, and guide your audience through your visualizations. Here are some tips for using captions and annotations effectively:

- **Be concise:** Keep your captions and annotations brief and to the point. Lengthy text can be overwhelming and detract from the impact of your visualization.

- **Be strategic:** Use captions and annotations to emphasize key points and help guide your audience's attention. Avoid cluttering your visualization with too many annotations or captions.

- **Use consistent formatting:** Ensure your captions and annotations have a consistent style, font, and size throughout your visualizations. This will make your presentation appear more polished and professional.

Structuring Your Presentation

How you structure your presentation can significantly impact its effectiveness. Here are some guidelines to help you create a compelling narrative with your visualizations:

- **Start with an introduction:** Begin your presentation with a brief introduction that sets the stage for your data story. This could include an overview of the data you're working with, the goals of your analysis, and any key questions you want to answer.

- **Organize your visualizations logically:** Arrange your visualizations in a way that guides your audience through your narrative. You might start by presenting an overview of the data, then drilling down into more specific insights, and finally offering conclusions or recommendations.

- **Provide context:** As you present your visualizations, provide context to help your audience understand the significance of the data. Explain why certain trends or patterns are important and what they might mean for the topic at hand.

- **End with a conclusion:** Conclude your presentation with a summary of your key findings and any implications or recommendations that emerge from your analysis. This reinforces your message and leaves a lasting impression on your audience.

Integrating storytelling into your data visualization process will allow you to transform your visualizations into compelling narratives that engage and inform your audience. As you practice this skill, you will become more adept at crafting data stories that captivate, inspire, and drive informed decision-making.

Design Considerations and Accessibility

Here, we will explore the fundamental design considerations and accessibility principles for effective data visualization. This encompasses aspects like color choices, font sizes, and layout, as well as ensuring that your visualizations are accessible and comprehensible to all viewers, regardless of any disabilities they might have.

Color Choices

Color can have a profound impact on the readability and interpretability of your visualizations. Here are a few principles to guide your use of color:

- **Use color to highlight and differentiate:** Color can draw attention to key elements in your visualization and distinguish different categories or data series. However, it's essential to ensure your colors are not too similar, which could lead to confusion.

- **Consider color connotations:** Colors often carry meanings or emotions, which can vary by culture. For example, red often signifies danger or loss, while green might denote growth or positivity. Be aware of these associations when selecting your colors.

- **Ensure sufficient contrast:** Make sure there is enough contrast between your text and background colors. This ensures that your text is easily readable against its background.

- **Be mindful of color blindness:** Approximately 8% of men and 0.5% of women of Northern European descent have a form of color blindness. Use colorblind-friendly palettes or tools that simulate color blindness to ensure your visuals are accessible.

Font Choices

Your choice of font can also significantly impact your visualization. Here's what you need to consider:

- **Choose a readable font:** Select a font that is easy to read, even at small sizes. Serif fonts can be easier to read in print, while sans-serif fonts often work better on screens.
- **Maintain consistency:** Use the same font type throughout your visualization to maintain a cohesive look. Using too many different fonts can make your visualization look cluttered and confusing.
- **Use size to show hierarchy:** Use font size to indicate the importance of text elements. Larger text should be used for titles and key points, while smaller text can be used for less important details.

Layout Considerations

The way you arrange elements in your visualization can affect its clarity and effectiveness. Here are some layout principles:

- **Place important information at the top or center:** Viewers typically look at the top or center of a visualization first, so place your most important information there.
- **Use white space wisely:** White space, or empty space, can help prevent your visualization from appearing cluttered and can guide the viewer's eye.
- **Align elements for clarity:** Align elements such as text, charts, and images to create clear lines and borders. This makes your visualization easier to understand and looks more professional.

Accessibility Principles

Making your visualizations accessible is vital to ensure that everyone, including people with disabilities, can understand your data. Here are some guidelines:

- **Use alt text:** Alt text provides a textual description of a visual element, which can be read out loud by screen readers for those with visual impairments. This text should concisely describe the content and purpose of the visual.

- **Avoid relying solely on color:** For those with color vision deficiencies, distinguishing colors can be challenging. Avoid conveying information through color alone. Instead, use patterns, labels, or other distinguishing markers.

- **Ensure sufficient contrast:** Elements with low contrast can be difficult for individuals with visual impairments to discern. Make sure there is sufficient contrast between text and background colors and between different elements in your visualization.

Incorporating these design considerations and accessibility principles will allow you to create visualizations that are not only aesthetically pleasing but also effective and accessible. These guidelines will help ensure that your visualizations communicate your data story as clearly and inclusively as possible.

Advanced Visualization Techniques

In this section we'll delve into some of the more advanced visualization techniques. While the basic types of visualization like bar charts, line graphs, and scatter plots are incredibly useful and commonly used, there's a whole world of advanced visualization techniques that can provide new insights and perspectives on your data. We'll focus on three areas: using R for visualization, creating web-based interactive visualizations with JavaScript, and exploring the realms of virtual reality (VR) and augmented reality (AR) visualizations.

Using R for Visualization

R, a powerful programming language for statistical computing, is highly valued for its graphical capabilities. While it can be complex to learn initially, the investment in understanding R can pay off with increased flexibility and customization in your visualizations.

Here are some important R packages for data visualization:

- **ggplot2:** This is one of the most popular visualization packages in R, implementing a "grammar of graphics" approach. With ggplot2, you can create complex multi-layered visuals by adding different components.
- **plotly:** This package allows you to create interactive plots that you can zoom, pan, and hover to display tooltips with more information.
- **leaflet:** If you're dealing with geographic data, leaflet is a fantastic package that lets you create interactive maps.

Creating Web-based Interactive Visualizations with JavaScript

JavaScript provides a rich ecosystem of libraries for creating interactive web-based visualizations. Two popular libraries are:

- **D3.js:** D3 (or Data-Driven Documents) allows you to bind data to the DOM and apply data-driven transformations to the document. With D3, you can create anything from a simple bar chart to complex infographics and interactive applications.

- **Chart.js:** If D3 seems overwhelming, Chart.js might be a good alternative. It offers simpler, more straightforward plotting options but with less customization.

Virtual Reality (VR) and Augmented Reality (AR) Visualizations

VR and AR technologies are opening up new possibilities for data visualization:

- **VR visualizations:** VR offers a fully immersive experience where you can interact with your data in a three-dimensional space. This can be especially useful for visualizing complex multidimensional data.

- **AR visualizations:** With AR, you overlay data onto the real world, offering an interactive and immersive way to explore data. AR can be useful for visualizing spatial or geographic data, such as overlaying traffic patterns on a street view.

Remember, while advanced visualization techniques can be exciting and eye-catching, they should only be used when they add value to your data story. It's always key to ensure that any visualization you use effectively communicates the insights in your data. By exploring these advanced techniques, you open new avenues for expressing the narratives hidden within your data, pushing beyond the limitations of traditional two-dimensional charts and graphs.

Evaluation of Visualizations

Now we'll delve into the process of evaluating your visualizations. Creating effective visualizations is one thing, but ensuring they effectively communicate your intended message is another. Here, we'll cover key criteria to consider when evaluating your visualizations, approaches for getting feedback, and the importance of iterative design in visualization.

Key Evaluation Criteria

There are several crucial elements you should assess when evaluating your data visualizations:

- **Clarity:** The most critical aspect of any visualization is its ability to convey the intended message clearly. Ask yourself, "Does this visualization effectively communicate the key insights?" Misinterpretation can lead to misguided decisions, so clarity should always be a top priority.

- **Simplicity:** While complex visualizations might look impressive, they can often confuse the viewer. Simplicity is often the best approach. "Does this visualization present the information as simply as possible, without losing any key insights?"

- **Accuracy:** A visualization should accurately represent the underlying data. Misrepresentations can lead to flawed interpretations and decisions. "Does this visualization accurately reflect the data?"

- **Relevance:** Not all data points are relevant to your message. Including extraneous information can distract from your key points. "Does this visualization focus on the most relevant data?"

- **Aesthetics:** While it shouldn't overshadow the other criteria, an aesthetically pleasing visualization can engage the viewer and enhance their understanding. "Is this visualization visually appealing?"

Approaches for Getting Feedback

It can be difficult to evaluate your own visualizations objectively, making feedback from others invaluable. Here are a few ways to gather that feedback:

- **Peer review:** Have colleagues or others in your field review your visualization. They can provide professional input and catch any inaccuracies or misconceptions.
- **User testing:** If possible, conduct user testing sessions, observing how individuals interact with your visualization and asking them to describe their interpretations.
- **Surveys and questionnaires:** These can be useful for gathering quantitative feedback on specific aspects of your visualization, like its clarity or aesthetic appeal.

The Importance of Iterative Design

Creating a good visualization rarely happens on the first try. Instead, you should approach it as an iterative process:

- **Design:** Create a draft of your visualization, focusing on representing the data accurately and clearly.
- **Evaluate:** Use the criteria above to evaluate your visualization. Solicit feedback from others, if possible.
- **Revise:** Based on your evaluation, revise your visualization. This might involve simplifying the design, emphasizing different data points, or adjusting the colors for better contrast.
- **Repeat:** Continue this process of design, evaluation, and revision until your visualization effectively communicates your intended message.

Evaluating your visualizations is a critical part of the data visualization process. By following these guidelines, you can ensure that your visualizations aren't just attractive, but they serve their ultimate purpose: to accurately and effectively communicate the insights within your data.

Module 5 Exam Questions

1. What is the main principle of data visualization?

a) To display data in a complicated manner

b) To create the most intricate visuals

c) To simplify complex data and allow for easier interpretation

d) To use as many colors as possible

2. Which of these is a widely used tool for creating data visualizations?

a) PowerPoint

b) Microsoft Word

c) Tableau

d) Notepad

3. What is a bar chart best used for?

a) Comparing different categories of data

b) Showing a trend over time

c) Depicting geographic distribution

d) Displaying correlations between variables

4. What is the primary reason for considering color blindness when creating data visualizations?

a) It affects the design aesthetics

b) It is required by law

c) To make visualizations accessible to all viewers

d) All of the above

5. Which font type is often better for reading on screens?

a) Serif fonts

b) Sans-serif fonts

c) Gothic fonts

d) Decorative fonts

6. Why is white space important in a data visualization?

a) It highlights important information

b) It guides the viewer's eye

c) It prevents the visualization from appearing cluttered

d) All of the above

7. What is an essential aspect of data storytelling?

a) Making the story complex

b) Using as many data points as possible

c) Highlighting key insights and weaving them into a cohesive narrative

d) None of the above

8. What is one benefit of using R for data visualization?

a) It allows for customization

b) It's easier to use than other languages

c) It doesn't require coding

d) It is specifically designed for creating bar charts

9. D3.js and Chart.js are libraries used for creating visualizations in which programming language?

a) Python

b) C++

c) Java

d) JavaScript

10. Which of these is an example of an advanced visualization technique?

a) Creating interactive web-based visualizations

b) Creating VR and AR visualizations

c) Using advanced R packages for data visualization

d) All of the above

11. Why should visualizations be as simple as possible?

a) Complex visuals are always difficult to understand

b) Simple visuals are quicker to create

c) To avoid confusing the viewer and ensure clarity of message

d) None of the above

12. What does it mean if a visualization is said to be accurate?

a) It has high-quality graphics

b) It includes all possible data points

c) It accurately represents the underlying data

d) It uses a variety of colors

13. What is the goal of user testing during visualization evaluation?

a) To test the user's knowledge

b) To see how individuals interact with the visualization and interpret the data

c) To demonstrate how the visualization should be used

d) None of the above

14. Why is it important to have a design iteration process when creating visualizations?

a) To include as many design elements as possible

b) To improve the visualization based on feedback and evaluation

c) Because it is a requirement for all data visualization projects

d) None of the above

15. What is geospatial data visualization?

a) Visualization that displays trends over time

b) Visualization that uses geographic data to display information on maps

c) Visualization that shows data distribution on bar graphs

d) Visualization that illustrates connections in social networks

16. Which of these tools allows for the creation of interactive dashboards and reports?

a) Microsoft Excel

b) Adobe Photoshop

c) Notepad++

d) Tableau

17. Which technique is particularly beneficial when working with large datasets, due to its ability to reduce storage space and improve processing speed?

a) Data Encoding

b) Data Decompression

c) Data Compression

d) Data Encryption

18. Which of the following visualization tools is primarily used for geospatial data visualization?

a) PowerBI

b) Tableau

c) ArcGIS

d) Excel

19. What aspect of your data visualization does the 'accessibility' part of design considerations primarily focus on?

a) Making sure the colors match the company branding

b) Making sure the design is visually impressive

c) Making sure the visualization can be understood by everyone, including people with disabilities

d) Making sure the visualization is printable

20. What is the main goal of evaluating your visualizations?

a) To find out if they are visually appealing

b) To ensure they effectively communicate the intended message

c) To make sure they can be understood by experts only

d) To check if they are compatible with all computer systems

Answers:

1. c
2. c
3. a
4. c
5. b
6. d
7. c
8. a
9. d
10. d
11. c
12. c
13. b
14. b
15. b
16. d
17. c
18. c
19. c
20. b

Module 6. Data Privacy and Security in Big Data

Introduction to Data Privacy and Security in Big Data

Welcome to Module 6 on Data Privacy and Security in Big Data. In this module, you will delve deep into the significant and complex landscape of privacy and security, specifically within the context of Big Data. As you will discover, the safeguarding of information is a pivotal responsibility when handling substantial volumes of data, and it is a cornerstone of a trustworthy data management strategy.

As we continue to generate an unprecedented amount of data every day, there is an increasing demand for systems and strategies that can protect data from unauthorized access, ensure the privacy of individuals and organizations, and comply with various regulations. Businesses, governments, and individuals are constantly seeking better ways to protect their valuable information while still being able to utilize it effectively. This module will act as a guide to understanding the intricacies of privacy and security within the realm of Big Data.

Throughout this module, we will explore the compelling intersection of Big Data and data privacy, scrutinize the importance of data security, and contemplate the ethical implications and legal frameworks surrounding the collection, storage, and analysis of large data sets.

To bring these concepts to life, we'll illustrate with practical examples from different industries, such as healthcare where maintaining patient privacy while utilizing data for better diagnostics is a significant challenge, or in finance where protecting sensitive financial information is a continuous battle against cyber threats.

Additionally, we'll also discuss how big tech companies, often seen as custodians of large volumes of personal data, are managing the delicate balance of harnessing the power of data analytics while maintaining the trust of their user base. For example, a key aspect of our discussion will be about the high-profile data breaches and the aftermath on both corporate reputation and legislative action.

Data privacy and security is a dynamic field, constantly evolving with technological advancements, regulatory changes, and growing public awareness. By the end of this module, you will have a solid understanding of the challenges and the mechanisms used to address these issues in the world of Big Data.

Join us as we navigate the intricate and crucial path of Data Privacy and Security in Big Data. This journey will provide invaluable insights for those seeking to ensure data protection while leveraging the true potential of Big Data in various domains. Let's dive in!

Data Protection Regulations

As you step into the world of Big Data, one of the first things you must understand and adhere to is the data protection regulations. These are sets of rules and guidelines designed to protect the privacy and security of personal data. With the increase in data breaches and cyber threats, the role of data protection regulations has become more crucial than ever.

Understanding these regulations is not just about compliance; it's about securing your data, protecting the rights of the data subjects, and building trust with your customers, clients, or users. Let's take a detailed look at some of these regulations and how they might affect you and your work with Big Data.

General Data Protection Regulation (GDPR)

Enacted by the European Union (EU) in 2018, the General Data Protection Regulation (GDPR) has had significant global impacts on data privacy and protection. Even if you are not based in the EU, it's essential to understand that GDPR applies to all organizations that process the personal data of people living in the EU, no matter where your organization is located.

GDPR aims to give control back to the individuals over their personal data. It also simplifies the regulatory environment for international businesses by unifying regulations within the EU. Under GDPR, you are required to obtain explicit consent from users to collect and process their data. The users also have a right to access their data, correct it, delete it ("right to be forgotten"), and transfer it to other service providers.

Non-compliance with GDPR can lead to severe penalties, up to €20 million or 4% of your global annual turnover of the preceding financial year, whichever is higher. It's important to note that the GDPR is more than just a checklist of things to do; it requires you to demonstrate that you've implemented data protection principles into your data processing activities.

California Consumer Privacy Act (CCPA)

For those operating in the United States, or handling the data of California residents, the California Consumer Privacy Act (CCPA) comes into play. Like GDPR, the CCPA provides California residents with increased control over their personal data, granting them the right to know what personal data is being collected about them, whether their personal data is sold or disclosed, and to whom.

If you fall under CCPA, you must respect these rights and provide methods for consumers to submit requests to exercise them. Non-compliance could result in penalties of up to $7,500 per intentional violation.

Brazil's General Personal Data Protection Law (LGPD)

If you work with data from Brazil's residents, you must comply with the LGPD. This law gives individuals similar rights as the GDPR and CCPA, such as access, correction, anonymization, portability, and deletion of personal data.

The LGPD also mandates the appointment of a Data Protection Officer (DPO) for organizations processing a large volume of personal data. The penalties for non-compliance can go up to 2% of your company's revenue in Brazil, up to a maximum of 50 million reais per violation.

Now that we've explored some prominent data protection regulations, it's essential to realize that each country may have its own set of data protection laws and regulations. Depending on the nature of your business and the markets you operate in, you may be subject to additional or different regulations. It's crucial to familiarize yourself with and adhere to these laws to ensure data privacy and avoid substantial penalties.

In the next section, we will explore more ways to protect your data and maintain compliance, focusing on techniques such as anonymization and pseudonymization.

Anonymization and Pseudonymization Techniques

The role you play in data privacy doesn't end with adherence to regulatory standards. As a custodian of Big Data, it's your responsibility to implement measures that ensure the privacy of the individuals whose data you manage. Two such strategies are anonymization and pseudonymization. These techniques allow you to use and analyze data while minimizing the risk of identification of individuals.

Anonymization

Anonymization is a process that removes or modifies identifiable information from a dataset in such a way that the individuals whom the data describe remain anonymous. When data is fully anonymized, it is not possible to re-identify the individuals, even by the data collector.

Consider a dataset of patient records. To anonymize it, you would remove or alter all identifiable information, such as names, social security numbers, addresses, and anything else that might lead back to the individual patient. In essence, after anonymization, the data should be free of any identifiable or unique entries that can be traced back to a person.

Several techniques can be used for data anonymization, including:

- **Data masking or obfuscation:** In this method, identifiable data is hidden or replaced with random characters or values. For instance, you might change all Social Security Numbers in your dataset to a series of X's.

- **Generalization:** This technique reduces the precision of data to make re-identification more difficult. For example, you might replace exact ages with age ranges, or specific locations with broader ones.

- **Data swapping or shuffling:** Here, values are interchanged between records, making it harder to link specific data to an individual.

- **Noise addition:** This involves adding random data ("noise") to existing data to obscure the original values.

Remember, anonymized data must be truly non-identifiable. If any data can be linked to an individual, directly or indirectly, it's not fully anonymized.

Pseudonymization

While anonymization completely removes identifiable information, pseudonymization replaces private identifiers with pseudonyms, or artificial identifiers. This way, data can still be attributed to a specific individual with an additional key. Pseudonymization reduces the linkage between the dataset and the individual's identity to protect it from data breaches or hacking attempts.

Imagine you're managing a dataset containing customer feedback for your products. To pseudonymize this dataset, you might replace each customer's name and contact details with a unique code. You maintain a separate, secure lookup table that links these codes back to the actual individuals. In this way, you can still use the feedback data for analysis and, if necessary, trace it back to specific customers.

However, if the pseudonymized data were to fall into the wrong hands without the key, those individuals would not be identifiable.

Pseudonymization techniques include:

- **Cryptography based methods:** These involve encrypting data using a secure key. The original data can only be restored if the key is available.

- **Tokenization:** This method replaces sensitive data with non-sensitive equivalents, called tokens, that have no meaningful value on their own.

- **Data blurring:** This technique involves slightly altering data values.

Remember, while pseudonymization can enhance privacy, it is not as secure as anonymization. Pseudonymized data can potentially be re-identified if the additional key is available, making it fall within the scope of data protection laws like GDPR.

Now that you understand these techniques, you might wonder: "Which should I use?" The answer depends on your needs. Anonymization provides a higher level of privacy and can take data outside the scope of data protection laws, but it also makes the data less useful, especially for personalized uses. Pseudonymization, on the other hand, retains more usability, but is less private and remains within the scope of data protection laws.

In the end, your role in managing Big Data requires you to walk a delicate line between utilizing data for insights and protecting the privacy of individuals. With a solid understanding of data protection regulations and privacy-enhancing techniques like anonymization and pseudonymization, you can navigate this path effectively and responsibly. As we continue our journey, we'll explore more techniques to secure our data, such as data encryption and access control.

Data Encryption

When it comes to the safety of your Big Data, encryption is one of the most reliable and widely used methods for securing data both in transit and at rest. Encryption is the process of encoding data in such a way that only authorized parties can access it. In essence, encryption turns readable data (plaintext) into coded form (ciphertext), which is not easily understood if intercepted.

Now, let's break down the two primary forms of encryption: symmetric and asymmetric encryption.

Symmetric Encryption

In symmetric encryption, the same key is used to both encrypt and decrypt the data. When you want to secure a file, you use the key to encrypt it. If you or someone else needs to access the file, you use the same key to decrypt it. The challenge with symmetric encryption lies in securely managing the key since anyone who has the key can decrypt the data.

Consider you have a large dataset containing sensitive customer information. To keep this data secure, you might use a symmetric encryption algorithm like AES (Advanced Encryption Standard) to encrypt the data. You must then keep the encryption key safe and secret. If anyone wants to access this data, they would need the encryption key to decrypt it.

Asymmetric Encryption

Unlike symmetric encryption, asymmetric encryption uses two different but mathematically linked keys: one for encryption (public key) and one for decryption (private key). You can share the public key with anyone, but you keep the private key secret. When someone wants to send you a secure message, they can encrypt it with your public key, and then only your private key can decrypt it.

Imagine you're receiving sensitive data from a number of sources. You could distribute your public key to those sources. They would then use that key to encrypt the data before sending it to you, and you would use your private key to decrypt the data upon receipt. This allows for secure data transmission, even over insecure networks.

Asymmetric encryption is the backbone of many secure protocols, including SSL and TLS, which secure the data transfer between web browsers and servers.

While asymmetric encryption is more secure, it's also slower and requires more computational power. Therefore, it's common to use a hybrid approach in practice: asymmetric encryption to securely share a symmetric key, then symmetric encryption to secure the actual data.

In your journey with Big Data, encryption is an essential tool. But remember, encryption only secures your data from being understood if it's accessed without authorization. Other strategies are needed to control access to the data in the first place, which brings us to our next topic: access control and auditing. As you continue to enhance your understanding of data privacy and security, keep in mind that the most effective strategies often involve a combination of methods, tailored to the specific needs and risks of your data.

Access Control and Auditing

While encryption plays a critical role in securing data, controlling who can access the data and tracking who accessed what are also vital aspects of a comprehensive data protection strategy. These concepts are encapsulated in the practices of access control and auditing.

Access Control

Access control is a method of guaranteeing that users are who they say they are (authentication) and that they have the appropriate permissions to access and perform operations on a given set of data (authorization). The primary goal of access control is to minimize the risk of unauthorized access to your data.

Access control can be managed in various ways:

- **Role-Based Access Control (RBAC):** Under RBAC, permissions are associated with roles, and users are assigned to these roles. For example, in a Big Data project, you might have roles such as data analyst, data scientist, and database administrator, each with different levels of access to your datasets. If a user is assigned the data analyst role, they will receive the permissions associated with that role.

- **Attribute-Based Access Control (ABAC):** ABAC controls access based on user attributes, environment conditions, and resource attributes. It's more flexible and granular than RBAC. For example, you could define a policy that only allows users in the 'Finance' department to access a certain financial dataset between 9 am and 6 pm.

- **Mandatory Access Control (MAC):** MAC restricts access to data based on the sensitivity of the information (labeled or classified) and the user's security clearance. It's often used in military systems or systems handling highly sensitive data.

Auditing

Auditing is about recording and examining activities in your system. By maintaining a comprehensive audit log, you can track who accessed what data, when they accessed it, and what operations they performed on it. Auditing is a powerful tool for detecting unauthorized access or potential data breaches, and for maintaining compliance with data protection regulations.

For instance, suppose a suspicious operation is performed on one of your databases. If you have auditing enabled, you could consult your audit logs to find out who performed the operation and when. If the action was unauthorized, you would have valuable information to investigate further and possibly prevent similar incidents in the future.

Many Big Data technologies, such as Hadoop and NoSQL databases, come with built-in tools for access control and auditing. For instance, Apache Ranger is a popular open-source tool used to manage access control policies and audit logs in Hadoop environments.

However, setting up and managing access control and auditing can be a complex task. It requires careful planning and an understanding of your data and the needs and rights of your users. It also involves ongoing maintenance, as roles, permissions, and user needs may change over time.

As you move forward in your journey with Big Data, keep in mind that data privacy and security aren't optional or additional features; they are foundational elements that you must consider at every step. From understanding data protection regulations to implementing encryption, access control, and auditing, each layer adds to the safety and integrity of your Big Data environment.

Incident Response and Disaster Recovery

As crucial as it is to proactively manage the privacy and security of your Big Data, it's equally important to prepare for unexpected incidents that could compromise your data. This is where incident response and disaster recovery come into play. These strategies ensure that when a data breach or loss occurs, you're prepared to respond effectively, minimize the damage, and quickly restore operations.

Incident Response

Incident response is the methodology for handling security incidents, breaches, and cyber threats. A well-defined incident response plan allows you to effectively identify, minimize the damage, and reduce the cost of a data breach, while finding and fixing the issue that caused it.

The steps involved in an effective incident response are typically:

- **Preparation:** This is your chance to get ready before an incident occurs. You should have a documented incident response plan and a trained response team. You should also have tools in place to detect and analyze potential security incidents.

- **Identification:** When an incident occurs, it's critical to quickly identify it so that the response can begin. Detection tools and techniques, as well as a culture of security awareness amongst your team, are important for identifying potential incidents early.

- **Containment:** Once an incident is identified, it's important to contain it to prevent further damage. This might involve isolating affected systems or taking them offline temporarily.

- **Investigation and Eradication:** Here you investigate how the breach occurred, remove the cause of the breach, and ensure threats are completely removed from your systems.
- **Recovery:** This step involves restoring systems back to normal operations and confirming they are no longer compromised.
- **Lessons Learned:** After an incident is dealt with, a post-incident review should be conducted to learn from the incident and improve future response efforts.

Disaster Recovery

While incident response is geared towards security breaches, disaster recovery focuses on restoring your Big Data operations following a significant disruption, which could be caused by natural disasters, hardware failures, or other unforeseen problems.

Key elements of a disaster recovery plan include:

- **Data Backups:** Regularly backup your data and test your backups to ensure they can be restored if needed. Your backups should be stored securely in a separate location.
- **Recovery Procedures:** Document the steps to recover each component of your Big Data infrastructure. This includes restoring data from backups, but also involves setting up hardware, installing and configuring software, and more.
- **Fault Tolerance and Redundancy:** These are design principles that involve setting up your systems in a way that they can continue to operate, even when certain components fail. This might involve duplicating certain components or data, or spreading your data and operations across multiple locations.

For example, imagine you're working with a cloud-based Big Data platform and there's a major service outage in your primary cloud region. If you've designed your systems with redundancy, you could have your data replicated in a different region, allowing you to switch operations to that region and continue your work with minimal disruption.

Whether you're dealing with a security breach or a service disruption, being prepared with an incident response and disaster recovery plan is key to minimizing damage and downtime. Such readiness not only helps maintain your Big Data operations, but also builds trust with the individuals and entities whose data you're handling. As you move forward, remember that the security of Big Data isn't just about preventing incidents, but also about being prepared to respond effectively when they occur.

Privacy by Design Principles in Big Data

The idea of privacy as an afterthought in system design has given way to a more modern and proactive approach: Privacy by Design. This principle argues that privacy should be a fundamental component, integrated into the very architecture of systems, processes, and practices dealing with data. This concept is particularly relevant when you're dealing with Big Data, given its inherent complexity and potential privacy implications.

Privacy by Design consists of seven foundational principles:

- **Proactive not Reactive; Preventative not Remedial:** The Privacy by Design approach doesn't wait for privacy risks to materialize, nor does it offer remedies for resolving privacy infractions once they have occurred. It aims to prevent them from happening in the first place. For example, when setting up a new Big Data process, you should consider potential privacy risks and implement safeguards to prevent them, such as data minimization techniques or secure default settings.

- **Privacy as the Default Setting:** You should ensure that personal data is automatically protected in any given IT system or business practice. If an individual does nothing, their privacy still remains intact. For instance, if you're building a user-facing application that collects personal data, privacy-preserving settings should be enabled by default.

- **Privacy Embedded into Design:** Privacy should be embedded into the design and architecture of IT systems and business practices. It should not be bolted on as an add-on. This involves considering privacy throughout the entire engineering process. For example, if you're implementing a new machine learning model in your Big

Data environment, privacy considerations should be taken into account from the very start, not only during the data collection phase but also while processing and storing the data.

- **Full Functionality – Positive-Sum, not Zero-Sum:** Privacy by Design seeks to accommodate all legitimate interests and objectives in a positive-sum manner, not a zero-sum (win/lose) way. It avoids the pretense of false dichotomies, such as privacy versus security, demonstrating that it is possible to have both. For instance, implementing strong encryption doesn't mean sacrificing system performance. Modern encryption algorithms can run efficiently and maintain data privacy without significantly impacting system performance.

- **End-to-End Security – Lifecycle Protection:** This involves applying a strong security measure throughout the entire lifecycle of the data, from collection to processing to disposal. For instance, if you're storing data in a distributed file system, you should ensure the data is secured while at rest, in transit, and during processing, and that it's safely deleted or anonymized when no longer needed.

- **Visibility and Transparency:** Privacy by Design seeks to assure all stakeholders that whatever the business practice or technology involved, it is operating according to the stated promises and objectives, subject to independent verification. If you're using third-party services to handle data, you should verify their privacy practices and ensure that users are fully informed about how their data is being handled.

- **Respect for User Privacy:** Above all, you should respect the interests of the individual by offering strong privacy defaults, appropriate notice, and empowering user-friendly options. For example, users should be given the option to opt out of data collection or processing activities, and this should be a straightforward and accessible process.

Embedding these principles into your Big Data operations doesn't just improve data privacy; it can also lead to better overall system design, enhancing user trust and potentially improving regulatory compliance. This way, you're not just building a system that handles Big Data but one that respects user privacy and promotes data stewardship. As you continue to develop and evolve your Big Data systems, always consider privacy as a fundamental, integral part of the design process.

Module 6 Exam Questions

1. What is the first step in an effective incident response?

a) Identification

b) Investigation and Eradication

c) Preparation

d) Containment

2. Which of the following is NOT a Privacy by Design principle?

a) Privacy as the Default Setting

b) Full Functionality – Positive-Sum, not Zero-Sum

c) Privacy is Optional

d) Proactive not Reactive; Preventative not Remedial

3. What is one of the key elements in a disaster recovery plan?

a) Data Backups

b) Incident Response Team

c) Encryption Keys

d) Anonymization Techniques

4. What does GDPR stand for?

a) Global Data Protection Regulation

b) General Data Protection Regulation

c) General Data Processing Regulation

d) Global Data Processing Regulation

5. What is a key feature of pseudonymization?

a) It completely removes all identifying information

b) It replaces identifying fields within a data record

c) It encrypts all data

d) It deletes specific data

6. What method involves replacing identifying data with artificial identifiers?

a) Encryption

b) Anonymization

c) Tokenization

d) Pseudonymization

7. Symmetric encryption uses:

a) Different keys for encryption and decryption

b) The same key for encryption and decryption

c) A public key for encryption and a private key for decryption

d) None of the above

8. The process of verifying changes and movement of data through its lifecycle is known as:

a) Data Encryption

b) Data Governance

c) Data Lineage

d) Data Lake

9. Which regulatory act concerns data privacy for individuals within the European Union?

a) HIPAA

b) SOX

c) PCI-DSS

d) GDPR

10. In the context of Big Data, access control means:

a) Restricting physical access to data centers

b) Managing who has access to certain data sets and operations

c) Password protecting your data

d) Installing security cameras

11. What does CCPA stand for?

a) Comprehensive California Privacy Act

b) California Customer Protection Act

c) California Consumer Privacy Act

d) Customer Comprehensive Privacy Act

12. What is a primary goal of auditing in Big Data security?

a) To correct errors in data

b) To track data access and usage

c) To improve data analysis results

d) To increase data storage capacity

13. Which Privacy by Design principle involves offering strong privacy defaults and empowering user-friendly options?

a) End-to-End Security – Lifecycle Protection

b) Full Functionality – Positive-Sum, not Zero-Sum

c) Respect for User Privacy

d) Proactive not Reactive; Preventative not Remedial

14. What is one of the primary objectives of the Privacy by Design approach?

a) To react quickly to privacy breaches

b) To minimize the damage of privacy infractions once they have occurred

c) To prevent privacy infractions from occurring in the first place

d) All of the above

15. Which encryption method uses a pair of keys – one public, one private?

a) Symmetric Encryption

b) Asymmetric Encryption

c) Hash Function

d) None of the above

16. Fault tolerance and redundancy are principles that are primarily part of:

a) Incident Response

b) Data Protection Regulations

c) Disaster Recovery

d) Data Visualization

17. The practice of removing all identifying information from data to protect personal privacy is known as:

a) Pseudonymization

b) Encryption

c) Anonymization

d) Tokenization

18. What is the purpose of a Business Continuity Plan?

a) To ensure that business operations can continue in the event of a disaster

b) To prevent security breaches from happening

c) To manage data storage and access

d) To comply with GDPR

19. Which Privacy by Design principle ensures that all business practices and technologies are operating according to the stated promises?

a) Visibility and Transparency

b) Privacy as the Default Setting

c) End-to-End Security – Lifecycle Protection

d) Full Functionality – Positive-Sum, not Zero-Sum

20. What is the main difference between anonymization and pseudonymization?

a) Anonymization removes all identifying data, pseudonymization replaces it

b) Anonymization replaces identifying data, pseudonymization removes it

c) Anonymization encrypts data, pseudonymization does not

d) There is no difference

Answers:

1. c
2. c
3. a
4. b
5. b
6. d
7. b
8. c
9. d
10. b
11. c
12. b
13. c
14. c
15. b
16. c
17. c
18. a
19. a
20. a

Module 7. Big Data Governance

Introduction to Big Data Governance

In this module, you will embark on the crucial journey of understanding and mastering the discipline of Big Data Governance. Our journey into the realm of Big Data would be incomplete without a comprehensive exploration of the frameworks, practices, and techniques that ensure the proper management and utilization of vast datasets. Governance is not merely a regulatory necessity; it is a strategic instrument that fosters trust, ensures data quality, and facilitates effective decision-making.

The scale and complexity of big data present unique challenges and opportunities, requiring a governance strategy that is robust and adaptive. Big data governance goes beyond the traditional scope of data governance by addressing not only data quality and data management but also the ethical use of data, metadata management, data privacy, and data security. It involves the implementation of policies, procedures, and standards that manage data availability, usability, integrity, and security in an organization.

Picture a multinational corporation that collects petabytes of data daily from a myriad of sources, including IoT devices, customer interactions, transactions, and social media. The successful extraction of meaningful insights from this voluminous data necessitates meticulous organization and governance. Failing to do so can result in missed opportunities, misguided business strategies, and even legal issues.

This module will guide you through the intricacies of Big Data Governance with a clear, practical approach. It is designed to equip you with the necessary tools and best practices to effectively govern big data in any organizational setting. From learning about the governance framework that provides the guidelines for decision-making to data quality management that ensures the reliability of insights derived, you will get a holistic understanding of the subject.

We will also delve into the practical aspects of metadata management, a critical component of big data governance that aids in understanding what the data represents, its origins, and its relationship with other data. The concept of data lineage and provenance, which ensures transparency and trustworthiness in data, will also be addressed.

Whether you're a data professional seeking to broaden your understanding of big data governance or a business leader wanting to leverage data for strategic decisions, this module has something invaluable to offer.

Get ready to unravel the world of Big Data Governance, a world where the effective management of data assets ensures a robust, successful, and data-driven organization. Let's get started!

Data Governance Framework

Understanding a data governance framework is pivotal for ensuring your data is trustworthy, secure, and effectively utilized. Such a framework outlines the principles, policies, and processes that guide data management within your organization, ensuring you maintain both the quality and integrity of your data.

Think of the data governance framework as the blueprint for your data management efforts. Just like an architect needs a blueprint to construct a building, you need a data governance framework to effectively manage your data.

Firstly, the governance framework will help you define who within your organization has the authority and responsibility for your data assets. You need to establish roles such as data owners, data stewards, and data custodians.

For example, imagine you're the head of a large company with numerous departments. You need a data governance framework to ensure that data ownership is clearly assigned. The sales data, for instance, could be the responsibility of the Sales Department Head. They will be the ones accountable for the accuracy, completeness, and security of the sales data.

However, merely defining these roles is not enough. You also need to make sure that everyone knows and understands these roles. Transparency about roles and responsibilities helps reduce confusion and ensures a smooth governance operation.

The data governance framework also helps you set up procedures for handling data. What should you do if there's a discrepancy in your data? What if there's a data breach? A well-defined framework gives you a roadmap to follow in such scenarios, thereby ensuring that data issues are addressed systematically and effectively.

Moreover, the framework guides your data policies. Policies such as data privacy, data quality, and data security are vital to managing your data efficiently. For instance, your data privacy policy could specify who can access your data, under what circumstances, and what they can or cannot do with the data.

An example here could be your policy for dealing with customer data. You may have a policy stating that customer data is only accessible to employees on a need-to-know basis, thereby helping to protect the privacy of your customers.

Finally, a data governance framework aids in establishing standards. Standards help ensure consistency in your data management practices. When your team follows the same standards for data entry, data processing, and data reporting, it reduces errors and makes your data more reliable.

Let's consider a practical example. Suppose your team is dealing with customer addresses. One team member may enter the address as "123 Fake Street," while another may enter it as "123 Fake St." Even though these represent the same address, the difference in formats can cause confusion and data errors. By setting a standard for address entries, you can mitigate such issues.

Building and implementing a data governance framework may seem like a daunting task, but the benefits are immense. With a robust data governance framework in place, you are better equipped to manage your data, ensure its quality and security, comply with regulations, and make data-driven decisions.

In the subsequent sections of this module, we will delve deeper into the components of the data governance framework, including data quality management, metadata management, and data lineage and provenance. We will equip you with the knowledge and tools to implement a successful data governance strategy, so keep reading and expanding your understanding of big data governance.

Data Quality Management

Data quality management (DQM) is a vital component of your data governance framework. Essentially, it encompasses all the practices, processes, and tools you use to ensure your data is accurate, complete, reliable, and timely. Without high-quality data, your decision-making could be flawed, and you risk losing trust in your data.

Let's start with why data quality is so critical. Imagine you're making a decision based on data that is incomplete or inaccurate. The outcome can be misinformed decisions that lead to inefficient operations, lost opportunities, and ultimately, impact your bottom line negatively. Data quality is the cornerstone of all data-driven activities and decisions, so it's crucial that you manage it effectively.

So, how can you manage your data quality? Here's where DQM comes into play. The first step is to define what 'quality data' means for you. This could vary based on your industry, your goals, and the specific use-cases you have for your data. A good starting point is to ensure your data is accurate, complete, consistent, timely, and reliable.

Accuracy is ensuring your data is correct and free from errors. For example, you might have a database of customer contact information. If the phone numbers or email addresses are incorrect, you won't be able to reach your customers effectively.

Completeness refers to ensuring that no essential data is missing. Suppose you're analyzing customer behavior on your e-commerce platform. If you have incomplete data about the customer's journey, your analysis may not fully reflect the customer's experience.

Consistency involves ensuring your data adheres to standard formats and conventions across all data sources and platforms. Let's say you're dealing with date formats. If one system records dates as MM-DD-YYYY and another as DD-MM-YYYY, this could lead to confusion and errors in data processing.

Timeliness implies that your data is up-to-date and available when needed. If you're tracking stock levels in your warehouse, outdated data could lead to stockouts or excess inventory.

Reliability means that your data consistently provides an accurate view of what it represents. If you conduct a customer survey, for instance, the same set of responses should give the same insights every time you analyze them.

After defining quality data, your next step is to implement strategies and tools to ensure this quality. This could involve data validation checks during data entry, routine data audits, and implementing data cleansing tools that can help identify and rectify errors in your data.

Suppose you have a form on your website where visitors can sign up for a newsletter. A validation rule could ensure that the email address field is not left blank and that it follows the correct email format. This can help prevent inaccurate or incomplete data from entering your database.

Moreover, you should also consider setting up a data quality dashboard. This dashboard can give you a snapshot of the current state of your data quality, helping you identify and address any issues promptly.

Remember, data quality management is not a one-time task. It's an ongoing process that requires continuous monitoring and improvement. This might seem challenging, but the effort is worthwhile because quality data is the foundation for informed decisions, streamlined operations, and successful data-driven strategies.

In the following sections, we will explore the rest of the components of the data governance framework, namely metadata management and data lineage and provenance. By the end of this module, you'll be equipped with a well-rounded understanding of big data governance.

Practical examples

Here are a few practical examples of Data Quality Management (DQM) in action:

Customer Data Validation:

Imagine you run an e-commerce platform. Customer data is essential for your business operations, from processing orders to delivering personalized marketing campaigns. To ensure the accuracy and completeness of this data, you could implement validation rules at the point of data collection. For instance, during user registration or checkout, ensure that email addresses are in a valid format, phone numbers contain the right number of digits, and all mandatory fields are filled out. This way, you ensure data quality right from the beginning.

Data Cleansing:

Suppose you've acquired a large dataset for a marketing campaign, but the dataset is riddled with duplicate entries, incorrect values, or inconsistent formatting. Running the campaign on such data would yield suboptimal results and could also harm your brand's reputation. In this scenario, you would use data cleansing techniques to identify and rectify the errors. Tools like OpenRefine or Talend can help you in this endeavor, offering features to standardize, de-duplicate, and correct your data.

Data Auditing:

Consider a healthcare provider that maintains extensive patient records. Due to the sensitive nature of this information, ensuring data integrity is crucial. A regular data audit could be scheduled to scan for discrepancies or anomalies in the data. For example, patients' age data shouldn't suddenly change significantly, or a patient's medical record shouldn't suddenly go missing. Any such inconsistencies could trigger alerts for further investigation and correction, thereby maintaining data reliability.

Establishing Data Quality Metrics and Dashboard:

Suppose you're a data manager for a large enterprise. Your role involves overseeing various data quality aspects across multiple departments. You could establish key data quality metrics (like data accuracy rate, data completeness rate, etc.) that align with your organization's goals. A data quality dashboard can then be set up to monitor these metrics in real-time. This way, you're proactively identifying potential data quality issues and can address them promptly, thereby avoiding any downstream impact.

Master Data Management (MDM):

Imagine you're operating a global supply chain network. You have suppliers, warehouses, and customers all around the world, each possibly maintaining their own versions of key data entities like product information, supplier details, etc. Without a unified view of this data, you could run into issues like overstocking, understocking, or even shipping the wrong items. An MDM solution would help ensure consistency of these key data entities across your network, thereby enhancing data quality and enabling smoother operations.

Each of these examples demonstrates how DQM practices can significantly enhance the quality of data, supporting more accurate analysis, decision-making, and overall business operations.

Metadata Management

Diving into another vital aspect of big data governance, we encounter metadata management. In simple terms, metadata is "data about data." It provides essential context, detailing when, why, how, and by whom a particular data was created, and what that data means. Effective metadata management is crucial to understanding your data and enhancing its usability and reliability.

Imagine you come across a digital photo with no information about it. Without knowing who took it, when it was taken, and what it represents, it's challenging to understand its value or relevance. But if you know this photo was taken by a professional photographer, at a significant event, and it portrays an essential moment, suddenly, the photo becomes meaningful and useful. That's what metadata does to your data.

Metadata can be broadly classified into three types: descriptive, structural, and administrative.

Descriptive metadata is information that helps you find and understand data. This includes data like titles, tags, and descriptions. Suppose you have a database of products. The product name, description, and tags can serve as the descriptive metadata, helping you understand what each product is.

Structural metadata provides insights into how complex datasets are organized and related. For example, if you have a database of invoices, structural metadata might detail how line items, invoice totals, and tax calculations are related within each invoice record.

Administrative metadata gives information about the source, ownership, and management of the data. For your customer database, administrative metadata could include who created each record, when it was created or last updated, and who is responsible for maintaining it.

Metadata management involves creating, storing, and maintaining your metadata. It starts with defining what metadata you need to capture. This largely depends on what you want to achieve with your data.

Next, consider how you'll store your metadata. Depending on your needs and the volume of your data, you might choose to store metadata directly within your database (in-line), in a separate but linked database (out-of-line), or even in a dedicated metadata repository.

To illustrate metadata management in action, let's take the example of managing a database for a library. Here, the descriptive metadata could be the titles, authors, and genres of books. Structural metadata might include information about how chapters and sections are organized within each book. Administrative metadata could cover when and by whom each book was added to the library.

By capturing and managing all this metadata, you not only understand your data better but also make it easier to find the right books, maintain your library efficiently, and even provide valuable insights to your library visitors.

However, metadata management is not a set-it-and-forget-it activity. It requires ongoing effort. You need to keep your metadata up-to-date and consistent with your data, particularly when your data changes. You may also need to adjust your metadata as your needs evolve.

In essence, metadata management is like keeping a detailed catalog of your data. It enables you to understand your data better, use it more effectively, and ensure its reliability. So, as you progress in your data governance journey, don't forget to give metadata the attention it deserves.

In the coming sections, we'll explore other components of the data governance framework, including data lineage and provenance. Each section will help you delve deeper into the world of big data governance and its practical applications.

Practical examples

Here are some practical examples of metadata management:

Digital Asset Management:

Let's say you're a graphic designer working with a vast array of digital assets like images, videos, and graphics. Each asset might have metadata such as the creator's name, creation date, file size, file type, and copyright information. Organizing this metadata allows you to easily locate, use, and manage your assets. For example, if you need to find all images created by a specific designer last month, having well-managed metadata enables you to quickly retrieve the required files.

Document Management in a Corporate Environment:

Suppose you're part of a large corporation with hundreds of internal documents - from HR policies to project reports. Metadata such as author, creation date, department, and document type can significantly improve document management. For instance, if you need to find all project reports authored by a specific employee within the last quarter, metadata management makes this task efficient and straightforward.

Healthcare Data Management:

In a healthcare setting, patient records carry immense importance. Each patient's record might have metadata like patient ID, doctor's ID, date of visit, diagnosis codes, treatment codes, and more. Effective metadata management not only helps with data retrieval but also with maintaining patient privacy and compliance with healthcare regulations. For example, if you need to audit all cases of a specific diagnosis treated by a particular doctor, well-structured metadata can facilitate this.

Website Management:

If you manage a website, metadata plays a crucial role in SEO and user navigation. Page titles, descriptions, and keyword tags form the metadata that search engines use to understand and rank your site. Moreover, metadata such as publication dates and author names helps users navigate your content. Well-managed metadata can enhance your website's visibility and user experience.

Research Data Management:

In a research context, datasets may come with metadata describing the study's methodology, data collection instruments, data codings, etc. This information is crucial for other researchers to understand, replicate, or build upon the study. In essence, metadata management here ensures the data's usability and reliability beyond the original research team.

Each of these examples demonstrates how metadata management can greatly enhance data organization, retrieval, and usability in various scenarios. It highlights the importance of taking a structured approach to capturing, storing, and maintaining metadata as part of your overall data governance strategy.

Data Lineage and Provenance

In the realm of data governance, another significant aspect that deserves your attention is data lineage and data provenance. These terms relate to the origin, lifecycle, and transformation of data within your systems. Understanding these aspects enables you to trace the journey of your data, which is pivotal for maintaining data integrity, managing changes, and troubleshooting issues.

Data lineage refers to the journey of data from its source to its current state. It outlines the path the data has traveled, what transformations it has undergone, and where it has been stored. It's like the "GPS tracking" for your data, offering a detailed map of its journey.

On the other hand, data provenance is closely related but slightly different. It offers a historical record of the data, detailing who has accessed it, how it's been used, and what changes have been made to it. In essence, it's the "audit trail" for your data.

Let's dissect how data lineage and data provenance play vital roles in your big data governance strategy.

Data Lineage

Data lineage provides transparency about how your data has been processed and transformed. This can be incredibly beneficial in multiple scenarios.

For instance, when you're troubleshooting a data issue, data lineage can help you identify where along the journey the problem has originated. Suppose you discover an error in your report. You can trace the data back to its source, inspecting each transformation to find where the error was introduced.

Similarly, if you're planning changes to your data architecture or data processing algorithms, data lineage can provide insights into how these changes might ripple through your systems.

For instance, imagine you're considering a change to your data cleansing algorithm. Understanding your data lineage helps you predict what datasets will be affected and how this might impact your downstream analysis and reports.

Data Provenance

Data provenance, on the other hand, helps ensure the integrity and trustworthiness of your data by keeping track of who has interacted with the data and how.

Suppose you notice an unexpected change in a key metric. Data provenance can help you identify who last accessed or modified the data, providing leads on what might have caused the change.

In contexts where data security and privacy are crucial, data provenance provides a way to audit data access and usage, helping ensure compliance with regulations and internal policies.

For example, if you're dealing with sensitive customer data, keeping a detailed record of data provenance helps ensure that only authorized users are accessing the data and that they're using it in compliant ways.

Implementing data lineage and data provenance might seem like a daunting task, given the volume and complexity of big data. However, various tools and techniques can help you automate this process, including data lineage tracking features in data integration tools, metadata management solutions, and specialized data governance software.

As you advance on your journey of big data governance, understanding and implementing data lineage and data provenance will become instrumental in maintaining the quality, integrity, and security of your data. With every step you take, remember that your goal is to make your data more transparent, trustworthy, and valuable.

In the next sections, we will discuss the remaining aspects of big data governance, further enhancing your understanding and skills in this crucial discipline.

Data Catalog

As you dive deeper into big data governance, you'll encounter the concept of a data catalog. A data catalog serves as a centralized repository where you can maintain a record of your datasets and their associated metadata. Just like a library catalog that indexes books, a data catalog indexes your data assets, making it easier for you to find and understand the data you need.

A well-structured data catalog can prove to be invaluable in numerous ways. Here's why:

- **Data Discovery:** A data catalog allows you to locate relevant datasets quickly. Without a catalog, finding the right data can be like searching for a needle in a haystack. However, with a catalog, you can easily search for datasets based on their metadata such as name, description, or tags.

- **Data Understanding:** A data catalog provides insights into what each dataset represents, its structure, and its context. For instance, if you come across a dataset named 'Sales_2022', a quick look at the catalog can provide additional information such as who created it, what it contains, and how it's structured.

- **Collaboration:** A data catalog promotes collaboration by making it easier to share data and metadata within your team or organization. For example, if you need to share customer demographic data with your marketing team, a data catalog enables the team to easily locate and understand the data.

- **Compliance:** A data catalog helps ensure compliance with data regulations by providing visibility into where sensitive data is stored and how it's used.

Building a data catalog may seem like a daunting task, especially when dealing with big data. However, you can start small by cataloging your most critical datasets first and then gradually expand your catalog over time. Furthermore, there are tools available that can automate much of the cataloging process.

For instance, imagine you are managing data for an e-commerce business. Your data assets might include customer data, product data, sales data, and more. To start building your data catalog, you can begin with the 'Customer' dataset. For each customer record, you can capture metadata like the record's source, when it was created, who created it, and what each field in the record means. Once you've cataloged this dataset, you can move on to the 'Product' dataset, and so on.

Over time, you'll have a detailed catalog of all your datasets, which will help you and your team to easily find and understand the data you need. It will also provide transparency into your data landscape, which is a key aspect of effective data governance.

Remember, a data catalog is a living document that needs to be updated as your data changes and grows. It might require some effort, but the benefits you gain in terms of data discoverability, understanding, and compliance make it a worthwhile investment.

In the next sections, we'll explore other aspects of big data governance, including data policies and standards, data stewardship, and change management. These topics will further enhance your understanding and skills in this critical discipline.

Data Policies and Standards

Data policies and standards form an essential pillar of your big data governance framework. They define the rules and norms that govern how you interact with your data, ensuring consistency, quality, and security. When you establish clear data policies and standards, you create a solid foundation for your data management efforts.

Let's delve into what data policies and standards encompass and why they're important.

Data Policies

Data policies are rules or guidelines that dictate how data should be collected, stored, managed, and used within your sphere of operations. These policies could relate to various aspects of your data, including but not limited to:

- **Data Collection:** Defining what data can be collected, how it should be collected, and who can collect it. For instance, your data collection policy might stipulate that personal customer data must be collected only with the customer's explicit consent.

- **Data Storage:** Setting rules for where data can be stored, how long it should be retained, and what security measures should be in place. For example, your data storage policy might mandate that all customer data should be encrypted and stored in a secure, cloud-based database.

- **Data Access:** Determining who can access different types of data, when they can access it, and what they can do with it. For instance, your data access policy might allow only certain members of your team to access sensitive customer data, and only for approved purposes.

Data Standards

While data policies outline the "what" and "why," data standards provide the "how." They specify the technical and procedural details of how your data policies should be implemented. These could include:

- **Naming Conventions:** Standards for how your datasets, fields, files, and other data elements should be named. For instance, you might have a standard that all dataset names should be in CamelCase and should accurately reflect the dataset's contents.

- **Data Formats:** Standards for how data should be formatted. For instance, you might specify that all dates should be in the "YYYY-MM-DD" format and all phone numbers should include the country code.

- **Data Classification:** Standards for categorizing your data based on its sensitivity, purpose, or other criteria. For example, you might classify data as "public," "internal," or "confidential."

Creating and enforcing data policies and standards is a proactive way to prevent data issues and inconsistencies. They provide a framework for everyone in your organization to understand what good data management looks like.

Consider a scenario where you are managing data for an international e-commerce company. You might establish a data policy that mandates all transaction data must be collected in a specific format, stored securely in a cloud database, and can only be accessed by authorized personnel. Your corresponding data standards might stipulate that transaction dates should be recorded in UTC format, transaction amounts should be in US dollars, and dataset names should follow a specific naming convention.

By having these data policies and standards in place, you can ensure that your data is consistently collected, stored, and accessed, making it easier for you to manage and use your data effectively.

In the next section, we'll dive into another essential part of your data governance framework: data stewardship. This will continue to expand your understanding of big data governance, equipping you with the knowledge to establish a robust and effective governance framework.

Data Stewardship

As you're deepening your understanding of big data governance, it's essential to address the role of data stewardship. Data stewards play a critical part in enforcing data policies, maintaining data standards, and promoting data quality within an organization. In essence, they're the guardians of your data, ensuring it's handled appropriately and used responsibly.

Data stewardship encompasses a range of responsibilities, which can be broadly classified into the following categories:

- **Data Quality Assurance:** As a data steward, you'd be responsible for ensuring the quality of your data. This involves checking data for errors or inconsistencies, cleaning and standardizing data, and implementing data validation processes. You'd work to prevent data issues at their source, as well as identifying and rectifying issues that do occur.

- **Metadata Management:** Data stewards often play a significant role in managing metadata. This might include creating and maintaining your data catalog, defining and implementing metadata standards, and ensuring that metadata is accurate and up to date.

- **Data Policy Enforcement:** Data stewards help enforce data policies and standards. This involves monitoring and auditing data usage to ensure compliance with internal policies and external regulations. As a data steward, you'd also be instrumental in promoting awareness of data policies and standards within your organization.

- **Data Access Management:** Part of your responsibility would include managing who can access what data. This means setting and administering data access controls, and monitoring data usage to prevent unauthorized access or misuse.

- **Data Advocacy:** As a data steward, you'd serve as an advocate for good data practices within your organization. This involves promoting data literacy, helping others understand the value of high-quality, well-governed data, and driving a culture that respects and leverages data.

For instance, let's imagine you're a data steward at a healthcare company that collects and analyzes patient data. Your day-to-day activities might include validating new patient data to ensure it meets quality standards, updating the data catalog with metadata for new datasets, monitoring data usage to ensure compliance with HIPAA and other regulations, administering data access controls to protect patient privacy, and training new staff on the company's data policies.

Being a data steward can be a challenging role due to the breadth and importance of the responsibilities. However, it's also a rewarding role that gives you a central position in your organization's data governance efforts. By ensuring the quality and integrity of your data, you'd be helping your organization make better decisions, improve operations, and meet regulatory obligations.

In the following sections, we'll delve deeper into change management in data governance and tools and technologies for data governance. These topics will further enhance your capabilities in big data governance, providing you with the comprehensive knowledge needed to effectively manage and govern your data assets.

Change Management in Data Governance

As you navigate your journey in big data governance, you'll soon encounter the dynamic nature of the data environment. The datasets you manage, the technologies you use, the policies you implement, and even the team you work with are all subject to change. Managing these changes effectively is a critical part of data governance, and this is where change management comes in.

Change management refers to the process of planning for, implementing, and managing change in an organization. In the context of data governance, it involves preparing for and managing changes related to your data and data management practices. Let's explore some of the key aspects of change management in data governance.

Understanding the Need for Change

Change often stems from a need or an opportunity. In data governance, the need for change could arise due to various factors such as new regulatory requirements, evolving business objectives, technological advancements, or data quality issues. Understanding the reason behind the change is the first step in managing it effectively.

Planning for Change

Once you've identified the need for change, the next step is planning. This involves defining what changes need to be made, who will be responsible for implementing them, and how they will be carried out. It also involves considering how the changes will impact your current data environment and what steps can be taken to mitigate any negative impacts.

Implementing Change

Implementation is where the planned changes are put into action. This might involve updating data policies, adopting new technologies, reorganizing your team, or changing the way you collect, store, or use data. During implementation, it's crucial to communicate with all affected parties and provide the necessary training or support.

Monitoring and Review

After changes have been implemented, they should be monitored to ensure they're having the desired effect. This involves tracking key metrics related to the change and reviewing progress regularly. If the changes are not yielding the expected results, further adjustments might be needed.

To provide a practical example, imagine you're managing data for a company that has decided to migrate its data storage to the cloud. This is a significant change that could have wide-ranging impacts on your data governance practices.

In managing this change, you might start by understanding the reasons for the migration (e.g., cost savings, scalability, accessibility). You'd then plan for the change by outlining what needs to be done (e.g., selecting a cloud provider, migrating data, setting up security measures), who will do it, and how it will be done. You'd also consider the potential impacts of the change and plan for how to mitigate them.

During implementation, you'd execute your plan, communicating with your team about the changes and providing necessary training. After the migration, you'd monitor key metrics (e.g., data accessibility, cost, security incidents) to ensure the change is delivering the expected benefits. If there are issues, you'd make further adjustments as necessary.

As you can see, change management in data governance is a multifaceted and critical process. It helps ensure that your data governance practices stay aligned with your evolving needs and objectives, and that changes are carried out smoothly and effectively.

Next, we'll explore another integral aspect of data governance: tools and technologies for data governance. This will further expand your knowledge and skills, empowering you to implement and manage a robust big data governance framework.

Tools and Technologies for Data Governance

In your journey towards mastering big data governance, understanding the landscape of tools and technologies available to support your data governance efforts is pivotal. These tools not only simplify the tasks associated with data governance but also make them more efficient and effective.

Here's an overview of the categories of tools you might find invaluable for your data governance initiatives:

Data Catalog Tools

Data catalog tools help you create and manage a central repository where metadata about your data assets is stored. They enable you to document essential information about your datasets, including their sources, structure, definitions, relationships, and usage rules.

For example, you may use a data catalog tool to document that a particular dataset contains customer transaction data, sourced from your e-commerce platform, and is updated daily. It may also contain details about the data quality rules that apply to this dataset.

Data Quality Tools

These tools allow you to assess, monitor, and improve the quality of your data. They provide functionalities for data profiling (analyzing data to understand its quality), data cleaning (correcting or removing incorrect, incomplete, or inconsistent data), and data monitoring (tracking data quality over time).

For instance, a data quality tool may alert you to inconsistencies in a dataset, such as dates in different formats or missing values in a required field. You can then use the tool to standardize the date format and fill in or remove records with missing values.

Metadata Management Tools

Metadata management tools support you in creating, maintaining, and using metadata. They can help you define and enforce metadata standards, automate metadata creation and updating, and enable easy searching and browsing of metadata.

A practical application could involve using a metadata management tool to automatically generate metadata for new datasets and ensure they adhere to your organization's naming conventions and data format standards.

Data Privacy and Security Tools

These tools help you protect your data from unauthorized access and use. They include encryption tools, access control tools, and data masking tools, which replace sensitive data with fictional yet realistic data for use in testing or development environments.

For instance, you might use a data privacy tool to anonymize a dataset containing customer data before sharing it with a third party for analysis, ensuring you comply with data protection regulations.

Data Governance Platforms

Data governance platforms are comprehensive tools that offer functionalities across multiple areas of data governance. They often include capabilities for data cataloging, data quality management, metadata management, data privacy and security, and more.

For example, you might use a data governance platform to manage all your data governance activities in one place, from documenting your datasets and monitoring their quality to setting and enforcing data policies and standards.

Implementing the right tools and technologies for data governance can greatly enhance your effectiveness in managing and governing your data. However, it's essential to choose tools that align with your specific needs, capabilities, and goals.

Module 7 Exam Questions

1. What is a crucial first step in managing change in data governance?

a) Implementing new technology

b) Understanding the need for change

c) Hiring new team members

d) Creating new policies

2. Which of the following tools would you use to anonymize a dataset?

a) Data catalog tool

b) Metadata management tool

c) Data privacy tool

d) Data governance platform

3. What is a primary purpose of a data catalog tool?

a) To correct data errors

b) To document essential information about datasets

c) To ensure data compliance with laws and regulations

d) To encrypt sensitive data

4. What is data stewardship?

a) The process of handling changes in data governance

b) The management of metadata for datasets

c) The enforcement of data policies and standards

d) The accountability for data quality, privacy, and compliance

5. Why is data lineage important in big data governance?

a) It enables data encryption

b) It helps to understand the journey of data through systems and transformations

c) It ensures data is stored in the cloud

d) It helps to track changes in data laws and regulations

6. What would you use data quality tools for?

a) To define and enforce metadata standards

b) To assess, monitor, and improve the quality of data

c) To document essential information about datasets

d) To protect your data from unauthorized access and use

7. In the context of big data governance, what is metadata?

a) The actual data you are analyzing

b) Data about the data, including its source, structure, and usage rules

c) Software used to analyze the data

d) The outcome of data analysis

8. What does change management in data governance involve?

a) Ignoring changes in data environment

b) Planning for, implementing, and managing change in an organization

c) Changing data at random

d) Focusing only on technological change

9. Which tool category allows you to automate metadata creation and updating?

a) Data catalog tools

b) Data quality tools

c) Metadata management tools

d) Data privacy and security tools

10. Why is understanding the need for change a crucial part of data governance?

a) It's important for maintaining data security

b) It's required by law

c) It helps you manage changes effectively and align with evolving needs

d) It's not important

11. What is a data governance framework?

a) A set of technologies used for managing big data

b) A blueprint for how to collect data

c) A structure outlining the processes, policies, standards, roles, and metrics that ensure effective data management

d) A tool for visualizing data

12. In the context of big data governance, what is data provenance?

a) The source and history of a piece of data

b) The accuracy and consistency of data

c) The security measures used to protect data

d) The policies governing data usage

13. What are data policies and standards used for?

a) To control the volume of data

b) To establish the rules for data usage and management in an organization

c) To predict future data trends

d) To visualize data

14. Which tool would be best suited for assessing data quality?

a) Data catalog tool

b) Metadata management tool

c) Data privacy tool

d) Data quality tool

15. What is the purpose of using data encryption tools in data governance?

a) To correct data errors

b) To protect sensitive data from unauthorized access

c) To document essential information about datasets

d) To manage metadata

16. How does a data governance platform aid in data governance?

a) It offers functionalities across multiple areas of data governance

b) It allows for data visualization

c) It helps in analyzing big data

d) It enables the creation of big data

17. Why is data quality management vital in big data governance?

a) It allows data visualization

b) It ensures the accuracy, consistency, and reliability of data

c) It helps in data encryption

d) It allows the creation of big data

18. What role does a data steward typically play in data governance?

a) They focus only on data encryption

b) They enforce data compliance with laws and regulations

c) They ignore changes in the data environment

d) They are accountable for data quality, privacy, and compliance

19. Which of these is not a responsibility of a data steward?

a) Ignoring changes in the data environment

b) Ensuring data quality

c) Protecting data privacy

d) Ensuring data compliance

20. Why would you use a metadata management tool in data governance?

a) To ensure data compliance with laws and regulations

b) To assess, monitor, and improve the quality of data

c) To document essential information about datasets

d) To create, maintain, and use metadata

Answers:

1. b
2. c
3. b
4. d
5. b
6. b
7. b
8. b
9. c
10. c
11. c
12. a
13. b
14. d
15. b
16. a
17. b
18. d
19. a
20. d

Module 8. Big Data Challenges and Future Trends

Introduction to Big Data Challenges and Future Trends

Welcome to Module 8 on Big Data Challenges and Future Trends. In this module, you will embark on an exploration of the core challenges and prospects that lie on the horizon of big data.

As we step into an era of exponential data growth, it is vital to understand that the field of big data isn't without its unique set of difficulties. Throughout this module, we will delve into the heart of these challenges, which range from scalability and performance issues to the complexities of data integration and interoperability. The solutions to these challenges will not only shape the future of big data but also define how businesses, governments, and institutions harness the power of data.

Our journey won't stop at just overcoming obstacles. We will turn our gaze forward and try to perceive the shape of things to come. You'll learn about the transformative potential of technologies like Edge Computing, the Internet of Things (IoT), and Quantum Computing, and how they promise to redefine big data and its applications.

For instance, consider the field of healthcare. Today, hospitals around the globe grapple with massive volumes of patient data. These include electronic health records, imaging data, genetic information, and data from wearable devices. The challenge lies in not just storing this vast ocean of data but also in ensuring its seamless integration and efficient analysis. How can we leverage Edge Computing and IoT to deliver real-time, personalized patient care, while still maintaining stringent data privacy standards?

On the other hand, Quantum Computing, while still in its infancy, is already showing great promise for the future. When mature, it has the potential to process complex data and perform calculations at speeds unfathomable with today's computers. But what will

Quantum Computing's impact on big data be? How will it transform our approaches to data storage, processing, and security?

These are just a few questions we'll grapple with in this module. Our aim is not only to equip you with knowledge but also to stimulate your curiosity, and encourage critical thinking about the path that lies ahead for big data.

By the end of this module, you will have a nuanced understanding of the current challenges in big data, an awareness of the future trends that are on the horizon, and the confidence to face these challenges head-on. Whether you are a data analyst, a manager overseeing a data team, or an enthusiast wanting to stay ahead of the curve, this module promises to deliver insights that will prepare you for the dynamic world of big data. Let's delve in!

Scalability and Performance

Big data's immense value is undeniable, but with this value come equally substantial challenges. Two of the most pressing are scalability and performance. Let's explore these in-depth.

Scalability refers to the system's ability to handle increasing amounts of workload effectively. In big data, scalability concerns how well your system or infrastructure can accommodate an expanding volume, variety, and velocity of data. Your system should be able to scale up or out, meaning it should be capable of increasing its computing power (scale-up) or increasing its number of systems (scale-out) as needed.

As a simple example, let's consider you're operating a popular social media platform. Initially, you may have a few hundred users generating data. However, as your platform grows and acquires millions of users, your initial system may struggle to store and process this much larger volume of data. That's where scalability comes in: your data storage and processing systems need to be able to grow in line with your platform. This could mean adding more storage space, more processing power, or both.

In contrast, performance is about how effectively your system can process data. It includes aspects like system speed, latency, and throughput. To understand it better, let's look at another example. Say you're running a real-time analytics operation for a large online retailer. During a big sale or a holiday season, the number of transactions will spike, and with it, the volume of data. If your system can't process this data in real-time, you might miss out on important insights, like sudden changes in customer behavior, that could be used to enhance your strategies during the sale.

The challenge of scalability and performance in big data can be daunting, but understanding these concepts is the first step to managing them effectively. Let's now look at some practical ways to address these issues.

When dealing with scalability, one key solution is to employ distributed systems, where you spread your data across multiple machines or clusters. Distributed systems can be scaled horizontally (more machines) or vertically (more powerful machines), allowing you to expand capacity as your data grows. Apache Hadoop and Apache Cassandra are excellent examples of technologies built with scalability in mind, as they can distribute data and computation across a large number of nodes.

As for performance, one of the most effective ways to improve it is by optimizing your data processing. This might involve using more efficient algorithms, optimizing your code, or even using hardware accelerators like GPUs. Additionally, you might want to consider using in-memory computation technologies like Apache Spark, which can significantly speed up processing times by storing data in the system's RAM rather than on disk.

Another method to enhance both scalability and performance is through the use of cloud solutions. Cloud service providers like Amazon Web Services, Google Cloud, or Microsoft Azure offer scalable storage and compute resources on demand, which can easily be adjusted to meet your changing needs. These services also provide a range of tools and applications designed to optimize performance, from data compression methods to machine learning algorithms.

In conclusion, addressing scalability and performance challenges in big data is a multifaceted task requiring a deep understanding of your data, the right tools, and ongoing management. But don't be daunted. The experience you gain by navigating these challenges will not only improve your data operations but also make you a more versatile data professional. As you learn to scale your systems and boost their performance, you'll be well-prepared to derive the most value from your data, regardless of its volume, velocity, or variety.

Data Integration and Interoperability

With the ever-growing volume and variety of data sources, two fundamental challenges arise in the field of big data: data integration and interoperability. Let's dive deep into what these terms mean and how you can tackle these challenges in a practical setting.

Data integration refers to the process of combining data from various sources into a unified view. In the context of big data, you might be dealing with a multitude of heterogeneous data sources, such as databases, data warehouses, flat files, APIs, social media feeds, and even IoT devices. The goal is to harmonize this data to allow for effective analysis and decision making.

Imagine you're a data scientist working for a multinational corporation. You may have sales data stored in a traditional SQL database, customer feedback coming in through social media, and inventory information on a cloud data warehouse. To extract valuable insights, you need to integrate these disparate sources of data to form a comprehensive view of your business.

On the other hand, interoperability refers to the ability of different systems, applications, or components to exchange and make use of information. It's about ensuring that your big data tools and technologies can interact seamlessly with each other and with other IT systems within your organization.

Think about a situation where you're using Apache Hadoop for storing and processing your big data, a NoSQL database for managing unstructured data, and a Business Intelligence tool like Tableau for data visualization. Interoperability ensures these disparate systems can communicate effectively, thereby enabling a smooth data flow from raw input to valuable insights.

Both data integration and interoperability are crucial for creating an efficient and reliable big data infrastructure. So, how can you navigate these challenges effectively? Let's delve into some practical approaches.

For data integration, the first step involves understanding the nature and structure of your data sources. What type of data are you dealing with? Is it structured or unstructured? What are the formats of the data? Once you've established an understanding, you can decide on the appropriate integration strategy. Techniques like Extract, Transform, Load (ETL) or Extract, Load, Transform (ELT) are commonly used. You may also consider data integration tools like Apache Nifi or Talend that offer out-of-the-box solutions for various integration scenarios.

When working with interoperability, a key consideration is the choice of technology. Always aim for tools and technologies that adhere to open standards and provide robust APIs. This will ensure that your chosen technology can interact effectively with other systems. For instance, if you're using a Hadoop-based system for your big data processing, you might want to ensure that it can interface well with your preferred data visualization tool.

Consider the implementation of middleware solutions, like Apache Kafka or RabbitMQ, that act as a messaging backbone, allowing different systems to communicate and exchange data efficiently. These technologies can greatly enhance your system's interoperability, promoting efficient data flow.

You may also want to consider Data Virtualization, which provides a way to access all your data across different sources, without needing to move it to a central repository. This can significantly ease your integration efforts and also enable a higher degree of interoperability.

Addressing data integration and interoperability issues in big data requires a combination of strategic planning, technological savvy, and a deep understanding of your data landscape. As you learn to integrate your data sources and ensure interoperability between your systems, you will unlock the full potential of your data, paving the way for better insights and more informed decision-making.

Edge Computing and IoT

In the realm of big data, two technologies have rapidly been gaining traction due to their profound impact on how we handle data: Edge Computing and the Internet of Things (IoT). In this section, let's explore what these technologies entail and how they intertwine to shape the landscape of big data.

Edge Computing is a computing paradigm that brings computation and data storage closer to the location where it's needed, aiming to improve response times and save bandwidth. In traditional cloud computing, data is sent from devices to a centralized data center or cloud for processing. With Edge Computing, however, processing takes place right at the edge of the network, often directly on the devices themselves or on a local server.

Consider this: you're operating a fleet of autonomous vehicles. Each vehicle is equipped with numerous sensors generating high volumes of data, including video feeds, location data, and performance metrics. Given the need for real-time decision-making, relying on a distant cloud server for processing this data could result in significant latency. By utilizing Edge Computing, you can process the crucial data directly on the vehicles, enabling immediate responses to changing road conditions, traffic, or potential hazards.

On the other hand, IoT refers to the network of physical objects or "things" embedded with sensors, software, and other technologies for the purpose of connecting and exchanging data with other devices and systems over the internet. These "things" can range from everyday household items like smart thermostats or refrigerators to complex industrial machinery.

To illustrate, assume you're managing an IoT-enabled smart home system. Your thermostat, lights, security cameras, and even your refrigerator are all connected, continuously generating data about their status, usage, and environment. By analyzing this data, you can derive useful insights and automate tasks, such as adjusting your home's temperature based on weather forecasts or time of day, or receiving a notification when your refrigerator is running low on groceries.

In the context of big data, Edge Computing and IoT go hand in hand. With the proliferation of IoT devices, there has been an exponential increase in the amount of data generated. If all of this data were to be sent to a central cloud for processing, it would result in substantial network congestion and latency issues. By integrating Edge Computing into IoT devices, you can address these challenges, processing large amounts of data locally and transmitting only the necessary data to the cloud.

In practice, the combination of Edge Computing and IoT can be seen in various sectors. In manufacturing, IoT sensors on factory equipment can detect anomalies in real-time, while Edge Computing enables immediate actions to prevent equipment failure. In healthcare, patient monitoring devices can gather vital data, and with Edge Computing, doctors can be alerted instantly in case of critical changes in the patient's condition.

Implementing Edge Computing and IoT in your big data strategy, however, is not without its challenges. You have to consider the architectural complexities of edge devices, security concerns, data privacy issues, and the interoperability of various IoT devices. However, with a clear understanding of these technologies and strategic planning, you can leverage Edge Computing and IoT to make your big data operations more efficient, responsive, and insightful.

Quantum Computing and Big Data

When exploring the future of big data, one cannot overlook the burgeoning field of quantum computing. This advanced technology has the potential to revolutionize how we process and analyze data. In this section, we'll delve into the relationship between quantum computing and big data, along with some practical examples to illustrate their synergies.

Quantum computing is a type of computation that harnesses the phenomena of quantum mechanics to process information. Traditional computers use bits for processing information, where each bit is either a 0 or a 1. Quantum computers, on the other hand, use quantum bits or qubits, which can be both 0 and 1 at the same time, thanks to a quantum property known as superposition. Another essential quantum property is entanglement, which allows qubits that are entangled to be linked together such that the state of one can instantaneously affect the state of the other, regardless of the distance separating them.

These unique properties provide quantum computers with a remarkable computational power. When dealing with problems involving big data, this increased power can significantly reduce processing times, opening new horizons for data analysis, machine learning, and artificial intelligence.

Imagine, for instance, you're tackling a complex optimization problem, like determining the most efficient delivery routes across a vast logistics network. A classical computer may need to check each possible route one by one, which could take an impractical amount of time as the number of routes grows exponentially with each additional location. However, a quantum computer, leveraging its ability to process multiple possibilities simultaneously, could find the most efficient route significantly faster.

Similarly, quantum computing could revolutionize machine learning, a field that's integral to big data analytics. Training complex machine learning models can be a resource-intensive process. Quantum computers, with their ability to perform multiple calculations simultaneously, could potentially train these models more efficiently, paving the way for more sophisticated models and accurate predictions.

Despite the enormous potential, it's important to remember that quantum computing is still in its early days, and practical, large-scale quantum computers are yet to become a reality. However, several tech giants, including IBM and Google, are actively researching and developing in this area, and quantum computing capabilities are increasingly being offered as a service over the cloud.

One such example is IBM's Quantum Experience, which allows you to experiment with quantum computing via the cloud. Here, you can explore and learn how quantum computing could potentially aid your big data endeavors, even before the technology becomes commercially ubiquitous.

Additionally, there are also potential implications on data security. Quantum computers could potentially break many of the existing cryptographic systems. As a big data professional, understanding these implications will be vital, and preparing for quantum-safe cryptographic methods could be a wise move.

In conclusion, while still emerging, quantum computing holds the promise of significantly impacting big data, potentially altering how you handle and analyze large datasets. Keeping abreast of developments in quantum computing and understanding its implications will help you stay ahead in the ever-evolving field of big data.

Data Quality in Big Data

In the big data world, data quality is one of the critical factors that can significantly influence your analysis and decision-making processes. Despite the exciting opportunities that come with the sheer volume, velocity, and variety of big data, it's of paramount importance that the quality of this data is maintained. In this section, we'll dissect what data quality means in the big data context, why it's important, and how you can ensure and improve it in practical terms.

Data quality can be characterized by several dimensions, such as accuracy, consistency, completeness, reliability, and timeliness. The accuracy of your data pertains to whether the data correctly represents the real-world values it's supposed to depict. Consistency refers to the need for data to be uniform across different data sources or over time. Completeness is about having no missing values in the dataset. Reliability is about the trustworthiness of your data source, while timeliness refers to how recent your data is.

Take a moment to think about a customer dataset you're managing. The accuracy could be compromised if a customer's name or address is recorded incorrectly. Consistency might be an issue if the customer's profile doesn't match across different systems in your organization. Your dataset is incomplete if essential data points, such as the customer's email address or phone number, are missing. The reliability might be questionable if you're not sure about the credibility of the source of your data. The timeliness of your data is compromised if it's outdated and doesn't reflect the current status of the customer.

Now, why is data quality crucial in big data? Simply put, poor data quality can lead to poor decisions. If you're basing your decision-making on inaccurate, inconsistent, or outdated data, it's highly likely that you'll draw incorrect conclusions and make suboptimal decisions. This can lead to financial losses, damaged customer relationships, and lost opportunities.

So, how can you ensure and improve data quality in your big data initiatives? Here are some practical steps you can take:

- **Data Profiling:** This involves assessing your data to understand its quality. By using data profiling tools or writing custom scripts, you can examine your data for inconsistencies, missing values, and other quality issues.

- **Data Cleansing:** This is the process of fixing or removing incorrect, corrupted, incorrectly formatted, duplicate, or incomplete data within a dataset. Techniques and tools for data cleansing can vary, from simple Excel functions to advanced software like OpenRefine or Trifacta.

- **Data Validation:** This involves setting up rules to check data for accuracy and consistency as it enters your system. You can enforce these rules using database constraints or custom validation logic in your data entry or integration processes.

- **Master Data Management (MDM):** MDM is a comprehensive approach to linking all critical data to a single point of reference. It aims to eliminate data inconsistencies and improve data quality across the organization.

- **Data Governance:** Implementing a robust data governance strategy can also help improve data quality. This involves defining who in your organization has authority and control over data assets and how these data assets may be used.

With high-quality data at your disposal, your analysis will be more accurate, your insights more reliable, and your decisions more informed.

Ethical Considerations in Big Data

In an era where data is widely considered as a valuable resource, it's imperative to consider the ethical implications of its use. As you navigate the world of big data, recognizing and adhering to ethical principles is crucial. In this section, we'll discuss the ethical dimensions of big data, such as privacy, consent, and fairness, and how you can ensure ethical practices in your big data endeavors.

Firstly, let's discuss privacy. As you handle large datasets, it's possible that you're dealing with sensitive information, whether it's personal data, financial details, or health records. Each piece of information has the potential to reveal intimate aspects of individuals' lives. Consequently, maintaining privacy becomes a significant ethical consideration. You need to ensure that data is anonymized or pseudonymized before analysis to prevent identification of individuals. Furthermore, you should employ robust security measures to prevent unauthorized access to your data, thereby protecting the privacy of individuals represented in your datasets.

Now, let's talk about consent. This pertains to obtaining permission from individuals before collecting, storing, and processing their data. In the context of big data, gaining explicit consent can be challenging due to the sheer volume and diversity of data sources. However, this doesn't diminish the ethical obligation to seek consent wherever possible. When using data from third-party sources, it's your responsibility to ensure that the data was collected with appropriate consent. For instance, if you're purchasing a dataset related to consumer behavior from a data broker, you must confirm that the data was gathered with the knowledge and consent of the individuals involved.

Fairness is another crucial aspect of big data ethics. With big data often used to drive decisions in areas like hiring, lending, or law enforcement, ensuring that your data and algorithms don't perpetuate bias or discrimination is essential. For example, if you're developing a machine learning model to automate your company's resume screening process, it's important to make sure your training data doesn't contain any inherent biases. This could mean checking that the data equally represents all genders, ethnicities, and age groups.

Lastly, let's touch on transparency. As you utilize big data, being open about your methods and intentions can help build trust. This means being clear about what data you're collecting, how it's being processed, and for what purpose. If you're using a customer's data to personalize their shopping experience, for instance, being upfront about it can increase their trust and engagement.

There are several ways you can put these ethical considerations into practice. Privacy-enhancing technologies, like differential privacy, can be employed to protect individuals' identities. Comprehensive data governance policies can be implemented to ensure consent, fairness, and transparency. Regular audits and ethical reviews of your data practices can also be helpful to maintain high ethical standards.

Remember, as a big data professional, you're not just a data handler, but also a data steward. It's your responsibility to ensure that your data practices respect individuals' rights and societal values. By doing so, you'll not only mitigate potential legal and reputational risks but also foster trust and generate long-term value from your big data initiatives.

Sustainability in Big Data

As we delve deeper into the digital age, the amount of data being generated and processed is growing exponentially. This massive data growth presents an often-overlooked challenge - sustainability. As you embark on your big data journey, understanding the environmental impact of data processing and storage, and how to mitigate it, is a critical consideration. In this section, we will explore the concept of sustainability in big data, why it matters, and how you can work towards more sustainable practices.

The rise of big data has led to an increased demand for data centers to store and process the data. These data centers consume significant amounts of energy, leading to substantial carbon emissions. For example, consider a popular streaming service like Netflix. Every movie you watch or song you stream is processed and delivered from a data center. These data centers, operating 24/7, use massive amounts of energy, contributing to the overall carbon footprint of the service.

So, how can you make your big data practices more sustainable? Let's delve into some practical steps.

Energy-Efficient Data Centers

When choosing a data center or cloud provider for storing and processing your big data, consider their energy efficiency and sustainability practices. Many leading data center providers are increasingly investing in energy-efficient infrastructure and renewable energy sources. By choosing such providers, you can contribute to reducing the environmental impact of your big data operations.

Data Minimization

The principle of data minimization involves collecting, storing, and processing only the data that is absolutely necessary. By minimizing the amount of data you work with, you can reduce the energy required for data processing and storage. For instance, if you're running a marketing campaign, you could focus on collecting only the key customer data necessary for the campaign, rather than amassing large amounts of unnecessary data.

Efficient Algorithms

The efficiency of your algorithms plays a significant role in the amount of computational resources required, and therefore the energy consumed. By designing and using more efficient algorithms, you can reduce the processing power needed, contributing to lower energy consumption. For instance, instead of using a brute force approach, you might use a more efficient sorting algorithm that reduces the number of operations needed to sort your data.

Green Computing Practices

Green computing practices aim to use computing resources efficiently while reducing environmental impact. This could include practices like virtualization, where multiple virtual systems run on a single physical system, reducing overall energy consumption. Another practice could be to optimize your data transfer to reduce energy use, for instance, by compressing data before transferring or scheduling transfers during off-peak hours to utilize energy more efficiently.

E-waste Management

Finally, managing electronic waste is another key aspect of sustainability. Outdated or broken hardware should be disposed of responsibly or ideally recycled. Many hardware providers offer recycling programs, and participating in these can help reduce the impact of electronic waste on the environment.

Making your big data practices more sustainable is not just about reducing environmental impact. It can also result in cost savings through reduced energy consumption and improved efficiency.

Moreover, demonstrating a commitment to sustainability can enhance your reputation and appeal to environmentally-conscious stakeholders. As you continue your journey in big data, remember to keep sustainability at the forefront of your considerations. Your choices can have a real impact on creating a more sustainable future.

Module 8 Exam Questions

1. What is the primary concern of the 'Scalability and Performance' in big data?

a) Data Visualization

b) Ability to process increasing volumes of data

c) Ethical considerations

d) Data privacy

2. What is the purpose of Data Integration in the context of big data?

a) To connect and streamline data from various sources

b) To visualize the data

c) To improve data storage solutions

d) To encrypt the data

3. Edge computing is closely related to:

a) Quantum computing

b) Data governance

c) Internet of Things (IoT)

d) Data lakes

4. Quantum computing can revolutionize big data by:

a) Enhancing data visualization

b) Achieving superior processing power

c) Reducing data storage costs

d) Improving data governance

5. In the context of big data, data quality refers to:

a) The size of the data

b) The speed at which data is processed

c) The accuracy, consistency, completeness, reliability, and timeliness of the data

d) The visual representation of data

6. Which is NOT a method to ensure data quality in big data?

a) Data profiling

b) Data compression

c) Data validation

d) Data cleansing

7. Ethical considerations in big data include:

a) Scalability and Performance

b) Privacy, consent, and fairness

c) Data integration and interoperability

d) Data visualization

8. What does transparency in big data ethics mean?

a) Using only open-source tools

b) Being open about your methods and intentions

c) Using clear visualizations

d) Always asking for explicit consent

9. Sustainability in big data is mainly concerned with:

a) Data governance

b) Mitigating environmental impact

c) Data integration

d) Scalability and performance

10. Energy-efficient data centers contribute to sustainability in big data by:

a) Increasing data processing speed

b) Reducing environmental impact

c) Enhancing data visualization

d) Improving data quality

11. The principle of data minimization involves:

a) Reducing the size of the data center

b) Reducing the number of data sources

c) Collecting, storing, and processing only the necessary data

d) Minimizing the use of data visualization tools

12. Efficient algorithms contribute to sustainability in big data by:

a) Reducing the need for data visualization

b) Reducing the energy consumption for data processing

c) Increasing data storage capacities

d) Improving data quality

13. What is a key aspect of green computing practices in the context of big data?

a) Using only green-colored visuals in data reports

b) Using computing resources efficiently while reducing environmental impact

c) Prioritizing renewable energy data in analytics

d) Always opting for physical data storage instead of cloud storage

14. The primary purpose of data governance in big data is to:

a) Improve data visualization

b) Ensure data quality and appropriate use of data

c) Speed up data processing

d) Reduce the cost of data storage

15. What is a significant challenge in ensuring data privacy in the context of big data?

a) Managing the high cost of data storage

b) Preventing unauthorized access to data

c) Speeding up data processing

d) Visualizing complex data

16. In the context of big data, consent refers to:

a) Agreeing to use open-source technologies

b) Obtaining permission from individuals before collecting, storing, and processing their data

c) Agreeing to use energy-efficient algorithms

d) Approving the use of certain data visualization tools

17. Which of the following is an essential ethical consideration in big data?

a) Using powerful processors

b) Increasing data storage capacity

c) Using proprietary software

d) Avoiding biases and discrimination in data and algorithms

18. What does 'edge computing' imply in big data processing?

a) Reducing data quality

b) Processing data closer to its source

c) Storing data in large data centers

d) Integrating data from various sources

19. Quantum computing is expected to impact big data by:

a) Reducing the need for data centers

b) Eliminating the need for data privacy

c) Solving complex problems more efficiently

d) Reducing the importance of data quality

20. In the context of sustainability in big data, e-waste management primarily involves:

a) Handling of outdated or broken hardware responsibly

b) Decreasing the use of electricity in data centers

c) Minimizing the amount of data processed

d) Eliminating the need for data visualization tools

Answers:

1. b
2. a
3. c
4. b
5. c
6. b
7. b
8. b
9. b
10. b
11. c
12. b
13. b
14. b
15. b
16. b
17. d
18. b
19. c
20. a

Module 9. Big Data Case Studies

Introduction to Big Data Case Studies

Welcome to Module 9 on Big Data Case Studies. In this module, we will dive into the real-world applications of big data, showcasing how it has revolutionized a multitude of sectors and industries.

Big data is not just a buzzword or a theoretical concept; it's a practical tool that is reshaping our world in profound ways. The way we work, communicate, consume, and even think is being influenced by the omnipresence of data. But the proof of the pudding is in the eating, and the best way to appreciate the transformative power of big data is to see it in action. This is precisely what we aim to do in this module.

Throughout this module, we will navigate through a variety of sectors and explore how big data is making a tangible difference. We will touch upon sectors such as healthcare, finance, retail, telecommunications, and urban planning for smart cities. Each of these industries presents unique challenges and opportunities for big data applications, and understanding these nuances is key to harnessing the full potential of big data.

For instance, in healthcare, big data is enabling predictive modeling to forecast disease outbreaks, enhancing patient care through personalized medicine, and optimizing hospital operations. In finance, big data aids in risk assessment, fraud detection, and customer segmentation. The list goes on.

We will unpack these case studies to reveal not just the power and potential of big data, but also the hurdles encountered and the lessons learned. Every sector has its own specific needs, rules, and restrictions. By observing how big data solutions have been adapted to fit these diverse scenarios, we can garner valuable insights into the versatility and adaptability of big data technologies.

In this module, you will learn how to translate the theoretical concepts and technical skills you've acquired into practical, impactful solutions. By studying real-world applications, you can better understand the potential of big data, its limitations, and how to creatively adapt to challenges.

This module is designed not just to inform but to inspire. As you explore these case studies, we hope that you'll start to envision how you can apply big data in your own field, contributing to the ongoing data revolution. These are not just stories of technology, but stories of innovation, resilience, and the transformative power of data. Let's get started!

Healthcare

In the healthcare sector, big data is revolutionizing how care is provided, monitored, and improved. As we delve into this sector, we'll explore how you can utilize big data to influence significant aspects of healthcare - from disease diagnosis to patient care, from public health to hospital administration.

Let's begin with an important use of big data: predictive modeling. With the enormous amounts of health data collected from sources like electronic health records (EHRs), genetic tests, and wearable fitness devices, you can build sophisticated models to predict disease outbreaks or identify potential health risks within a specific population. For instance, you could leverage machine learning algorithms to predict the probability of patients developing a chronic condition based on various factors such as age, family medical history, lifestyle habits, and genetic predispositions. This sort of proactive approach to healthcare, facilitated by big data, can lead to early interventions and improved patient outcomes.

Next, let's look at personalized medicine, one of the most promising areas of healthcare. Instead of a one-size-fits-all approach, personalized medicine uses big data to tailor treatments to your specific genetic makeup and health history. For example, by analyzing a patient's genetic data alongside their disease progression, you can identify specific biomarkers associated with their condition and customize their treatment plan accordingly. This individualized treatment can greatly improve treatment efficacy and minimize adverse effects.

Big data also shines in enhancing operational efficiency within healthcare settings. Hospitals often have to manage a multitude of factors - from bed availability, staff scheduling, to supply chain management. Big data tools can process real-time data from various hospital systems, helping you make data-driven decisions to optimize these operations. Imagine having a dashboard where you can monitor key operational metrics in real-time, allowing you to address any bottlenecks immediately.

Moreover, big data can greatly aid in medical research. The process of developing new drugs involves analyzing large volumes of scientific data, which can be streamlined using big data techniques. By effectively managing and analyzing the diverse data from clinical trials, you can accelerate the pace of drug discovery and development.

Public health is another area where big data can have an enormous impact. With big data, you can track and analyze health trends over large populations and geographies, helping guide public health policies and interventions. For instance, during a pandemic, real-time tracking and analysis of infection rates, hospitalizations, and mortality rates can provide critical insights that can help curb the spread of the disease.

However, it's essential to understand that with great potential comes great responsibility. Dealing with healthcare data involves navigating complex issues of privacy and security. Regulations like HIPAA (Health Insurance Portability and Accountability Act) in the United States provide guidelines for maintaining patient confidentiality while handling health data. As a big data professional in healthcare, you'll need to ensure compliance with such regulations and employ techniques like data anonymization and encryption to protect patient data.

Practical Examples of Big Data in Healthcare

Here are some practical examples of Big Data in Healthcare:

- **Predictive Analytics in Patient Care:** Using data collected from electronic health records (EHRs), wearables, and other data points, health systems can predict which patients are at risk for various diseases. For example, Kaiser Permanente uses predictive analytics to identify hospital patients at risk of sepsis, a severe and often deadly infection. By analyzing multiple data points from patient health records in real time, they can quickly identify at-risk patients and take immediate action.

- **Improving Medication Adherence:** CVS Health is using predictive analytics to improve medication adherence. By analyzing prescription data, patient behavior, and engagement in health programs, they predict which patients are likely to stray from their prescribed regimen. Interventions can then be designed to help these patients stay on track, improving their long-term health outcomes.

- **Precision Medicine:** The University of California, San Francisco (UCSF) uses big data in its cancer genomics project. They sequence tumor DNA to identify mutations that drive an individual's cancer. The data is then used to match patients with targeted treatments that are more likely to be effective given their specific genetic makeup, improving the overall success rates of cancer treatment.

- **Hospital Operations and Resource Management:** Johns Hopkins Hospital uses big data analytics to predict patient inflow during peak times. By analyzing EHRs, local event schedules, weather forecasts, and historical admission rates, they can optimize staff schedules and bed management, leading to a more efficient use of resources and improved patient care.

- **Public Health and Disease Surveillance:** During the COVID-19 pandemic, various public health organizations used big data to track the spread of the virus. The COVID-19 dashboard by Johns Hopkins University, which visualizes and tracks confirmed COVID-19 cases in real time, is a prime example. The dashboard uses big data gathered from various global health organizations to provide an up-to-date view of the pandemic.

Remember, while these examples show the power of big data in healthcare, it's also crucial to ensure the privacy and security of patient data when implementing such data-driven approaches.

In conclusion, the application of big data in healthcare is vast and exciting. It offers you the opportunity to significantly improve patient care, advance medical research, and streamline healthcare operations. In the next section, we'll look at how big data is influencing another dynamic sector - finance.

Finance

In the finance sector, big data offers profound implications for risk management, fraud detection, customer segmentation, and more. Let's navigate this complex terrain together to discover how you can harness the power of big data to transform financial operations and services.

First, consider risk assessment, a critical aspect of the finance industry. Traditionally, assessing a borrower's credit risk was based on a limited set of data points. With big data, you have access to a more comprehensive and nuanced view of a borrower's financial behavior. You can incorporate a wider range of data - from a client's transaction history, social media behavior, to economic trends - to build a more accurate credit risk profile. With this enriched perspective, you're better equipped to make lending decisions and manage your portfolio risk.

Let's also consider fraud detection. Financial fraud is a significant concern in the industry, and big data can serve as a potent weapon against it. By continuously monitoring and analyzing transactions in real-time, you can detect unusual patterns that might signal fraudulent activity. Machine learning models can be trained to recognize these patterns and alert you to potential fraud, allowing for swift action and reducing financial loss.

Another area where big data is making strides in the finance industry is customer segmentation. Big data analytics can offer deep insights into customer behavior, preferences, and needs. By analyzing data from various customer touchpoints, you can segment your customers into more refined categories. For instance, you can identify which customers are more likely to respond to a specific offer or which ones are at risk of churn. Such insights allow you to tailor your marketing strategies, improve customer engagement, and enhance customer retention.

Big data is also transforming investment strategies. In the world of high-frequency trading (HFT), even microseconds matter. Big data tools can help you analyze market trends and make trading decisions at lightning speed. Moreover, you can incorporate non-traditional data like news feeds, social media chatter, and even weather forecasts into your analysis to make more informed and timely investment decisions.

Beyond these applications, big data is revolutionizing financial planning and wealth management. By compiling and analyzing data from various sources, big data can offer you personalized financial advice. For instance, you can use predictive analytics to estimate future expenses, simulate different investment scenarios, and develop a tailored financial plan.

Lastly, in the regulatory compliance arena, big data plays a vital role. Financial institutions must comply with numerous regulations, and any violation can lead to hefty penalties. By monitoring and analyzing all your transactions and operations in real-time, big data can help you ensure that you're in compliance with all the necessary regulations. Moreover, in case of any audit or investigation, you can easily trace and present the required data.

As you navigate the finance sector with big data, remember that you're dealing with sensitive financial data, and it's crucial to uphold data privacy and security. Robust encryption and access control mechanisms need to be in place to safeguard the data.

Practical Examples of Big Data in Finance

Here are some practical examples of Big Data in Finance:

- **Risk Assessment:** JPMorgan Chase uses big data to predict the risk level of certain loans and credit products. By examining factors such as a customer's credit history, financial transactions, employment history, and social media behavior, they can more accurately predict the probability of loan repayment, therefore mitigating financial risks associated with lending.
- **Fraud Detection:** American Express uses machine learning and big data analytics to detect fraudulent

transactions. The system continually learns from the patterns it finds in the data. If a transaction seems suspicious (for instance, if it's made in a location far from the cardholder's typical area of use), the system will flag it for further investigation.

- **Customer Segmentation:** Citibank uses big data to segment its customers into more refined categories. By analyzing various customer touchpoints, such as spending habits, social media interactions, and responses to marketing campaigns, they can offer products and services that are more tailored to each customer's needs, thereby increasing customer satisfaction and loyalty.

- **Algorithmic Trading:** Many hedge funds and financial institutions use big data to make automated trading decisions. Renaissance Technologies, for instance, uses mathematical models and big data to predict price changes in financial markets. These models analyze vast amounts of market data, including real-time information, to make split-second trading decisions.

- **Regulatory Compliance:** Financial firms also use big data to ensure they are in compliance with regulatory requirements. For example, Goldman Sachs uses big data to track and analyze every trade they make. This allows them to quickly identify any transactions that might not comply with current regulations, helping them avoid potential fines and legal issues.

- **Personalized Financial Services:** Fintech companies like Mint or Personal Capital use big data to offer personalized financial advice to their users. They analyze a user's income, spending, saving habits, and financial goals to provide customized budgeting advice and investment strategies.

As with healthcare, ensuring the privacy and security of financial data is a critical aspect when working with big data in finance. Implementing robust data protection measures and adhering to financial data regulations are key considerations in these examples.

In summary, big data is transforming the financial landscape, allowing for smarter risk management, enhanced customer insights, efficient fraud detection, and more. As you move forward, keep exploring the possibilities that big data brings to the financial sector, because the possibilities are limitless. Up next, let's explore the application of big data in the retail sector.

Retail

Big data in retail offers a gold mine of opportunities for personalization, optimization, and innovation. Here, we'll go through how you can leverage big data to better understand customers, streamline operations, and stay ahead of market trends.

In retail, understanding your customer is key to success, and big data provides unprecedented insights into customer behavior. By analyzing data points from transactions, loyalty programs, online interactions, and social media, you can paint a detailed picture of your customer's preferences, habits, and needs. With this information, you can personalize your marketing efforts, sending the right message to the right customer at the right time. Imagine being able to predict what a customer is likely to buy next based on their past purchases and then sending them a tailored offer - this is the power of big data in retail.

Big data can also be instrumental in optimizing pricing strategies. Retail prices are not static and can be influenced by various factors such as demand, competition, and seasonal trends. By analyzing large datasets of historical sales, competitor pricing, and market conditions, you can optimize your pricing strategy to maximize sales and profitability. For instance, you can set up dynamic pricing models that automatically adjust prices based on real-time demand.

One of the biggest challenges in retail is inventory management. Overstocking leads to increased storage costs and potential waste, while understocking results in missed sales opportunities. Here, big data comes to the rescue by allowing you to accurately forecast demand. By considering factors such as historical sales, promotional calendars, and even weather forecasts, you can predict future sales and adjust your inventory levels accordingly.

In the e-commerce realm, big data can help you improve your website's user experience. By tracking and analyzing customer interactions on your website - from click patterns to browsing history - you can gain insights into how customers navigate your site and where they might encounter difficulties. You can use this information to make data-driven improvements to your website, making it easier for customers to find what they're looking for and increasing conversion rates.

Big data is also key to enhancing in-store experiences. By analyzing in-store video footage, you can understand customer foot traffic patterns and optimize store layout accordingly. Additionally, technologies like beacons can track customers' in-store movements, allowing you to send personalized in-store offers to their smartphones based on the aisles they're browsing.

However, just as in healthcare and finance, it's important to remember the responsibility that comes with handling customer data in retail. You must ensure that you're respecting customer privacy and complying with data protection regulations.

Practical Examples of Big Data in Retail

Here are some practical examples of big data in retail:

- **Personalization:** Amazon is a leader in using big data for personalization. They analyze data on customer behavior - including browsing history, past purchases, and items in the shopping cart - to recommend products that are most relevant to each customer. This personalized approach not only enhances the shopping experience but also drives additional sales.

- **Dynamic Pricing:** Uber uses big data to implement dynamic pricing, also known as surge pricing. Based on real-time supply and demand data, Uber adjusts its prices to balance driver availability with rider demand. This helps Uber ensure availability even during peak times.

- **Inventory Management:** Walmart uses big data to optimize inventory levels across its stores and warehouses. By analyzing sales data, Walmart can predict which items will be in high demand and ensure those items are well-stocked.

- **Customer Sentiment Analysis:** Starbucks uses big data to analyze customer feedback from various sources, including social media, customer surveys, and online reviews. This helps Starbucks understand customer sentiment and make improvements to their products and services.
- **Store Layout Optimization:** Many retailers, including Tesco, use big data to optimize store layouts. By analyzing customer footfall patterns and purchasing behavior, they can arrange products in a way that maximizes sales.
- **Supply Chain Optimization:** Companies like Zara use big data to streamline their supply chains. They analyze data on sales, returns, and customer demand to forecast trends and ensure their supply chain operations align with customer needs.

In all these cases, respecting customer privacy and adhering to data protection regulations is paramount. Also, it's important to remember that while big data can provide valuable insights, it's how you act on these insights that truly matters.

In conclusion, big data offers a wealth of opportunities for you to gain a competitive edge in the retail sector. It allows you to deeply understand your customers, streamline operations, and adapt quickly to changing market trends. As you explore big data in retail, keep asking yourself: How can I use these insights to better serve my customers? Up next, we'll delve into big data in the telecommunications industry.

Telecommunications

In telecommunications, big data presents a broad range of opportunities for enhancing network performance, improving customer experience, and enabling new business models. This segment will guide you through these opportunities and showcase how you can leverage big data to transform your operations in telecommunications.

Let's start with network performance. Networks generate vast amounts of data every second - data on calls, messages, data usage, and more. By analyzing this data, you can gain deep insights into network performance. For instance, you can identify patterns that signal potential network failures and take preemptive action. Also, you can optimize network traffic in real time based on usage patterns, ensuring the network runs smoothly even during peak times.

Big data also brings transformative possibilities to customer experience management. Through an analysis of data from call records, customer complaints, social media, and other touchpoints, you can understand the customer journey in depth. For example, you can identify the common issues customers face, predict which customers are likely to churn, and understand what factors drive customer satisfaction. With such insights, you can personalize your communication, resolve issues proactively, and develop services that truly resonate with your customers.

One important application of big data in telecommunications is in fraud detection. By analyzing call records and usage patterns in real time, you can detect anomalies that might indicate fraudulent activities, such as SIM card cloning or international revenue share fraud. Upon detecting potential fraud, you can promptly investigate and take action, protecting your revenue and your customers.

In the era of 5G and IoT, big data plays a crucial role in enabling new business models. For instance, you can leverage big data to offer data-driven services to other industries. An example of this is providing real-time traffic data to smart city initiatives or offering location-based insights to retailers.

Another exciting frontier for big data in telecommunications is predictive maintenance. By analyzing data from network equipment, you can predict when a piece of hardware is likely to fail and schedule maintenance before it affects network performance. This not only improves network reliability but also helps you manage maintenance costs by preventing emergency repairs.

Let's also not forget about revenue assurance. In the complex landscape of telecommunications billing, errors can easily creep in, leading to revenue leakage. By continuously monitoring and analyzing billing data, you can detect and rectify these errors, ensuring that you're billing accurately for the services provided.

As you delve into big data in telecommunications, remember the importance of data security and privacy. Protecting the personal data of your customers and complying with relevant data protection regulations should always be a top priority.

Practical Examples of Big Data in Telecommunications

here are some practical examples of big data in telecommunications:

- **Network Optimization:** Telecom operators such as Verizon use big data to monitor and optimize network performance. By analyzing data on network traffic, signal strength, and call drop rates, they can identify and resolve network issues, ensuring reliable service for their customers.

- **Customer Experience Management:** AT&T uses big data to enhance the customer experience. They analyze data from various customer touchpoints, including call records, customer service interactions, and social media, to understand customer behavior and preferences. This enables them to personalize communication and improve customer service.

- **Fraud Detection:** Vodafone uses big data for real-time fraud detection. By analyzing call patterns and usage data, they can detect anomalies that may indicate fraudulent activities, such as SIM card cloning or international revenue share fraud, and take immediate action.
- **Predictive Maintenance:** Telecom companies are increasingly using big data for predictive maintenance. By analyzing data from network equipment, they can predict when a piece of hardware is likely to fail and schedule maintenance before it leads to network downtime.
- **Churn Prediction:** Telecommunication companies like T-Mobile use big data to predict customer churn. They analyze data such as call records, billing information, and customer complaints to identify customers who are likely to switch to a different provider. This enables them to proactively address customer issues and retain customers.
- **Revenue Assurance:** Orange, a French telecom operator, uses big data to ensure accurate billing. They continuously monitor and analyze billing data to detect and rectify errors, preventing revenue leakage.

These examples illustrate the transformative power of big data in telecommunications.

In summary, big data opens up a wealth of opportunities in telecommunications, from network optimization and improved customer experience to fraud detection and new business models. As you explore these possibilities, continuously ask yourself: How can I use these insights to improve my services and offer more value to my customers? Next, we'll explore how big data is transforming the way we build and manage smart cities.

Smart Cities

The concept of Smart Cities is intrinsically tied to big data. In smart cities, every device, sensor, and system is connected, generating massive amounts of data that, when properly harnessed, can lead to more efficient, sustainable, and livable urban environments. In this segment, we will help you understand how you can leverage big data to transform cities into smarter, more connected communities.

Firstly, consider the domain of traffic and transportation. Traffic congestion is a common pain point in many cities. Here, big data can make a significant impact. By gathering real-time data from GPS devices, traffic cameras, public transportation systems, and even social media, you can develop dynamic traffic management systems. These systems can provide real-time traffic updates, optimize traffic signals to improve flow, and even predict traffic congestion before it happens, allowing city officials to take preventive measures.

Public transportation can also be greatly enhanced with big data. For example, by analyzing data from ticketing systems, GPS tracking, and passenger feedback, you can optimize bus routes and schedules, predict maintenance needs for transportation infrastructure, and enhance the overall public transit experience.

Big data also plays a crucial role in environmental sustainability. Smart grid technology, which leverages big data, allows for more efficient management of energy resources. For instance, real-time data on energy consumption can be used to balance supply and demand, reducing energy waste. Furthermore, data from weather systems and pollution sensors can be used to monitor and improve air quality.

Smart waste management is another application of big data in smart cities. Sensors in waste bins can send data on fill-levels, allowing waste collection routes to be optimized based on real-time needs rather than fixed schedules. This not only makes waste management more efficient but also reduces the carbon footprint of waste collection vehicles.

Public safety and security are also enhanced in a smart city through big data. Real-time data from surveillance cameras, emergency services, and social media can be analyzed to detect potential security threats or respond rapidly in case of emergencies. For example, predictive policing tools can analyze historical crime data to predict where and when crimes are more likely to occur, enabling more efficient allocation of law enforcement resources.

The key to creating a smart city lies in your ability to integrate and analyze data from various sources to draw actionable insights. However, as you collect and handle large amounts of data, it's vital to ensure the privacy and security of your citizens' information. Implement robust data protection measures and respect the laws and regulations regarding data privacy.

Practical Examples of Big Data in Smart Cities

Here are some practical examples of how big data is being used in the development of smart cities:

- **Traffic Management:** Google's Waze application collects data from users about traffic conditions, accidents, and other road incidents in real time. This data is then shared with other users and local municipalities to help optimize traffic flow and reduce congestion.

- **Public Transportation:** The city of Chicago uses big data to optimize its public transportation system. Data from GPS devices on buses and trains, along with passenger data, is used to adjust routes, manage schedules, and predict maintenance needs.

- **Waste Management:** In Seoul, South Korea, they use smart bins equipped with sensors to monitor waste levels. When a bin is about to be full, the sensor sends a signal to the waste management company, which can then plan its collection routes more efficiently.

- **Energy Management:** In the smart grid of Chattanooga, Tennessee, sensors across the city gather data on energy use in real time. This allows the city to balance energy supply with demand, identify power outages instantly, and even predict where future outages may occur.

- **Air Quality Monitoring:** Many cities around the world, including London and Beijing, use sensor networks to monitor air quality in real time. This data can then be used to inform the public about air quality conditions, guide environmental policy, and track the effectiveness of pollution-reduction measures.

- **Public Safety:** In New Orleans, the city uses big data to improve public safety. They analyze data from surveillance cameras, 911 calls, and social media to identify patterns and trends that can help predict where crimes are likely to occur.

- **Resource Management:** In Barcelona, they use big data and IoT technology for park irrigation. Sensors in the soil monitor moisture levels and switch the sprinklers on or off as needed, conserving water.

- **Urban Planning:** The city of Boston uses Waze data to identify areas of the city where there are consistent traffic problems. They use this data to inform urban planning decisions and infrastructure improvements.

These examples show the wide range of ways in which big data can be used to make cities smarter, more efficient, and more responsive to the needs of their residents. However, as with all uses of big data, it's essential to handle data responsibly and maintain the privacy of individual citizens.

In conclusion, big data is an integral part of building smart cities that are efficient, sustainable, and safe. As you explore these opportunities, remember to always focus on the ultimate goal: improving the quality of life for your citizens. As we wrap up this module, I hope you have gained valuable insights into the practical applications of big data across various sectors, and I encourage you to continue exploring and innovating with big data in your own field.

Agriculture

As you may already know, modern agriculture is increasingly driven by data. From precision farming to supply chain optimization, big data plays an integral role in feeding a growing global population while minimizing the environmental impact. In this section, we will delve into how you can utilize big data to revolutionize agricultural practices and ensure sustainable productivity.

Let's start with precision farming. With the use of sensors, drones, and GPS technology, you can collect detailed data about your fields. This data includes information about soil composition, moisture levels, nutrient needs, and pest presence. By analyzing this data, you can make more informed decisions about irrigation, fertilizer application, and pest management. Instead of treating your entire field the same, you can tailor your farming practices to the specific needs of different areas in your field. This not only improves crop yield and quality but also reduces waste and environmental impact.

Next, let's discuss yield prediction. Historical yield data, combined with weather data, crop genetic information, and soil data, can help you predict your future yield with high accuracy. Predictive analytics can provide insights into how changing weather patterns might affect your yield, enabling you to make proactive adjustments to your farming practices. Moreover, accurate yield predictions are crucial for managing your supply chain and market commitments.

Big data also plays a critical role in supply chain optimization. You can track your products from farm to fork, ensuring freshness, reducing waste, and enhancing accountability. For example, sensors can monitor the temperature and humidity conditions of your products during transport, alerting you if conditions fall outside the optimal range. Moreover, by analyzing data on market demand and logistics, you can optimize your distribution strategy to reduce costs and ensure your products reach the market at the right time.

Data can also be used for farm equipment maintenance. Many modern farming machines are equipped with sensors that collect data on machine performance and wear and tear. By analyzing this data, you can predict when a machine is likely to need maintenance, enabling you to schedule repairs before a machine breakdown occurs.

Finally, let's consider the role of big data in sustainable farming. Data on crop water needs and weather conditions can be used to optimize irrigation and conserve water. Meanwhile, data on the environmental impact of different farming practices can help you develop more sustainable methods. Additionally, satellite data can be used to monitor crop health and detect early signs of disease, reducing the need for chemical inputs.

Practical Examples of Big Data in Agriculture

Here are some practical examples of how big data is being applied in the field of agriculture:

- **Precision Farming:** John Deere, a leading agricultural machinery manufacturer, has been at the forefront of implementing precision farming. Their tractors and other machinery come equipped with sensors and GPS technology, which gather data about the soil and crops. This data is then used to apply water, fertilizers, and pesticides more efficiently, tailoring the application to the needs of specific areas within a field.

- **Yield Predictions:** The Climate Corporation, a digital agriculture company, provides farmers with a platform that integrates weather data, soil information, and field data to provide real-time monitoring and yield predictions. Farmers can use these insights to make more informed decisions about planting, irrigation, and harvesting.

- **Supply Chain Optimization:** IBM's Food Trust uses blockchain technology to create a traceable record of the journey of food products from farm to store. This not only ensures freshness and reduces waste but also improves accountability in the food supply chain.

- **Farm Equipment Maintenance:** Companies like AGCO use IoT technology to collect data on the performance of their farm machinery. This data is used to predict maintenance needs, allowing farmers to schedule repairs and prevent equipment breakdowns.

- **Sustainable Farming:** Semios, a precision farming platform, uses big data and IoT technology to help farmers reduce their chemical inputs. Sensors placed throughout a farm monitor pest populations, weather, and crop stress. Farmers receive alerts when action is needed, allowing them to apply pesticides only where and when they are needed.

- **Livestock Management:** Cowlar, a startup company, produces smart collars for cows. These collars collect data on temperature, activity, and behavior, which is then analyzed to provide farmers with insights into the health and productivity of their livestock.

These examples illustrate how big data is transforming agriculture, making farming more efficient, productive, and sustainable. The benefits are not only economic, improving yield and reducing costs, but also environmental, helping to conserve water, reduce chemical use, and minimize the impact of farming on the environment.

In conclusion, big data opens up vast possibilities for you in agriculture. It empowers you to make data-driven decisions that increase efficiency, productivity, and sustainability. As you explore the power of big data in agriculture, remember to handle your data responsibly and respect privacy considerations.

Manufacturing

The manufacturing sector, characterized by complex processes, intertwined supply chains, and constant needs for efficiency and innovation, presents an ideal scenario for the application of big data. Here, you will learn how you can leverage big data to revolutionize your manufacturing processes, boost efficiency, and innovate your product offerings.

The first stop is predictive maintenance. With Internet of Things (IoT) sensors placed strategically throughout your production facilities, you can continuously collect data on the condition of your machinery. This might include information about vibration, temperature, and wear and tear. By applying predictive analytics to this data, you can anticipate potential equipment failures before they occur, schedule maintenance accordingly, and prevent costly production downtime. This proactive approach can help you ensure continuous, smooth operations, and extend the lifespan of your machinery.

Next, consider the power of big data for quality control. In a traditional setup, quality checks are often carried out towards the end of the production line, which means that any flaws detected necessitate reworking or scrapping of the product. By contrast, with real-time data analytics, you can identify and address quality issues as soon as they arise during the production process. For example, machine learning algorithms can be trained to analyze image data from high-speed cameras on your assembly line, immediately detecting deviations from established quality parameters.

Moving forward, let's delve into demand forecasting. By analyzing historical sales data alongside data on market trends, economic indicators, and even weather patterns, you can better predict future demand for your products. This insight can inform your production planning, helping you optimize your inventory levels, reduce waste, and respond swiftly to market changes. Additionally, this improved forecast accuracy can also streamline your supply chain, ensuring that materials and resources are available exactly when they're needed.

Furthermore, big data can facilitate your efforts in process optimization. By analyzing operational data, you can identify bottlenecks, inefficiencies, or waste in your production process. For example, data on energy usage can highlight opportunities for reducing your energy consumption or shifting use to off-peak times to benefit from lower rates.

Lastly, let's examine the role of big data in product innovation. Through data analysis, you can gain deep insights into customer behavior, preferences, and needs. This data can fuel your product development process, leading to innovative products that resonate with your target audience. It could be as simple as identifying a common customer complaint in product reviews, which could spark a design modification, or as complex as noticing a trend in usage data that inspires a whole new product line.

Practical Examples of Big Data in Manufacturing

Here are some practical examples of big data being used in the manufacturing sector:

- **Predictive Maintenance:** General Electric's Predix platform is a prime example of big data application in predictive maintenance. The platform collects and analyzes data from machines and predicts future failures. This helps in scheduling maintenance activities, reducing downtime, and improving overall operational efficiency.

- **Quality Control:** Intel uses big data for real-time quality control in its manufacturing process. It applies predictive analytics to high-speed, high-volume data streams from the production line to detect anomalies and resolve issues as they occur, rather than at the end of the production process. This leads to fewer product defects, less waste, and significant cost savings.

- **Demand Forecasting:** Companies like Coca-Cola analyze big data from various sources including sales, marketing campaigns, and external factors like weather patterns to predict product demand. These insights help the company optimize its inventory levels and production schedules, leading to cost efficiency and better customer satisfaction.

- **Process Optimization:** Siemens uses big data in its manufacturing processes to optimize energy usage. It collects data from its machinery and production processes, then uses analytics to identify patterns and opportunities for reducing energy consumption and minimizing waste.

- **Product Innovation:** Procter & Gamble uses big data to analyze customer behavior and feedback from various sources, including social media and product reviews. This analysis feeds into its product development process, allowing the company to innovate and design products that resonate with consumers' needs and preferences.

- **Supply Chain Management:** Big data analytics play a key role in managing supply chains effectively. For example, IBM's Supply Chain Insights uses AI and big data to predict supply chain disruptions and suggest alternate suppliers and routes. This proactive approach helps to mitigate risks and ensures smooth operations.

These examples demonstrate the transformative power of big data in the manufacturing sector. By harnessing the potential of big data, manufacturers can improve operational efficiency, reduce costs, enhance product quality, and foster innovation.

In conclusion, big data can be a powerful tool for you in the manufacturing industry, enabling you to drive efficiency, maintain high quality, streamline operations, and foster innovation. As you venture into the world of big data, remember the importance of data security and privacy.

Module 9 Exam Questions

1. In the context of Big Data in agriculture, what is precision farming?

a) Predicting weather patterns

b) Broadcasting farming programs on the radio

c) Applying resources like water and fertilizers efficiently based on data

d) Monitoring the stock market

2. Which technology is commonly used in precision farming?

a) IoT and Big Data

b) Blockchain

c) Virtual Reality

d) Quantum computing

3. What is a common application of Big Data in sustainable farming?

a) Using data to optimize the use of chemical inputs

b) Building a social network for farmers

c) Using data to predict the winner of a farming competition

d) Creating a virtual reality farming game

4. Predictive maintenance in the manufacturing sector typically involves:

a) Predicting stock market trends

b) Predicting equipment failures before they occur

c) Predicting sales for the next quarter

d) Predicting the results of a football match

5. How is big data commonly used for quality control in manufacturing?

a) By analyzing high-speed, high-volume data streams from the production line to detect anomalies

b) By scanning barcodes on finished products

c) By tracking employee attendance

d) By monitoring social media posts about the company

6. What is the purpose of using Big Data for demand forecasting in manufacturing?

a) Predicting the best location for a new store

b) Predicting weather conditions

c) Predicting future demand for products

d) Forecasting the number of employees needed for a shift

7. In the field of healthcare, how can big data be used to improve patient care?

a) By predicting patient no-shows and cancellations

b) By optimizing the layout of the hospital cafeteria

c) By predicting trends in fashion and music

d) By tracking the sales of medical equipment

8. How does big data support disease monitoring and management in healthcare?

a) By analyzing the lyrics of popular songs

b) By predicting the result of political elections

c) By predicting stock market trends

d) By aggregating and analyzing patient data to track disease progression and response to treatment

9. In finance, what is one key application of big data?

a) Fraud detection and prevention

b) Predicting the winner of a football match

c) Forecasting weather patterns

d) Predicting the sales of a music album

10. How can big data improve customer experience in the retail industry?

a) By predicting future fashion trends

b) By predicting stock market trends

c) By personalizing customer interactions based on data about past purchases and browsing behavior

d) By monitoring the health of retail employees

11. How can big data improve operational efficiency in telecommunications?

a) By using data to optimize network performance and predict failures

b) By using data to decide the color of the company logo

c) By using data to predict future weather conditions

d) By using data to track employee attendance

12. In the context of smart cities, how can big data Improve transportation?

a) By predicting stock market trends

b) By using data to optimize traffic flow and reduce congestion

c) By predicting the results of sports events

d) By monitoring social media posts about the city

13. In agriculture, how can big data be used for supply chain optimization?

a) By creating a traceable record of the journey of food products from farm to store

b) By predicting the results of farming competitions

c) By monitoring the stock market

d) By forecasting weather patterns

14. In manufacturing, how does big data assist in product innovation?

a) By analyzing customer behavior and feedback for insights

b) By tracking employee attendance

c) By predicting political election results

d) By scanning barcodes on finished products

15. How can big data help in managing supply chains effectively in the manufacturing sector?

a) By predicting stock market trends

b) By predicting future demand for products

c) By predicting supply chain disruptions and suggesting alternate suppliers and routes

d) By monitoring social media posts about the company

16. What role does big data play in telecommunications network performance?

a) Predicting political election results

b) Predicting stock market trends

c) Predicting future fashion trends

d) Optimizing network performance and predicting failures

17. In smart cities, how is big data used in energy management?

a) By optimizing energy usage in buildings and across the city

b) By predicting the results of sports events

c) By tracking employee attendance

d) By predicting stock market trends

18. How can big data be used to improve public safety in smart cities?

a) By predicting stock market trends

b) By analyzing citywide data to predict and prevent crime

c) By predicting future fashion trends

d) By tracking employee attendance

19. In agriculture, how does big data assist in livestock management?

a) By collecting data on livestock health and productivity

b) By predicting the results of farming competitions

c) By predicting future weather conditions

d) By monitoring the stock market

20. In manufacturing, how is big data used to optimize energy usage?

a) By collecting data from machinery and production processes to identify opportunities for energy saving

b) By predicting future fashion trends

c) By monitoring social media posts about the company

d) By tracking employee attendance

Answers:

1. c
2. a
3. a
4. b
5. a
6. c
7. a
8. d
9. a
10. c
11. a
12. b
13. a
14. a
15. c
16. d
17. a
18. b
19. a
20. a

Module 10. Building a Career in Big Data

Introduction to Building a Big Data Career

Welcome to Module 10 on Building a Big Data Career. In this module, you will learn how to chart your career path in the dynamic, high-demand field of Big Data. This field continues to evolve and expand as the world generates data at an unprecedented rate and businesses are increasingly relying on data-driven insights to make strategic decisions.

We will navigate the vast landscape of career opportunities in Big Data, from roles such as Data Analysts, who interpret complex digital data for actionable strategies, to Data Scientists, who create predictive models and algorithms, to Data Engineers, who ensure the smooth flow and availability of data in an organization.

While roles vary, one thing remains constant: a successful career in Big Data requires a balance of technical know-how, analytical thinking, and communication skills. This module will provide you with a detailed understanding of the skills and qualifications you need, including which programming languages to master, how to use Big Data tools, and the soft skills that will make you an invaluable asset to any team.

Let's look at a practical example: imagine you're a marketing executive who has noticed a shift in customer behavior. Traditional analytical tools can't process the amount of data you have, so you're missing key insights that could shape your marketing strategy. Enter Big Data specialists, armed with the ability to parse this massive amount of data and derive valuable insights that can direct your marketing efforts and, ultimately, drive business success.

This module will also guide you through the critical steps of crafting an effective resume, preparing for interviews, and negotiating salaries. Furthermore, we will delve into strategies for continuous learning and staying up-to-date with industry trends in order to sustain a long-term career in Big Data.

We will hear real-world success stories and learn from failures in various Big Data careers through our case studies. These examples will provide you with a comprehensive understanding of what it's like to work in Big Data and will demonstrate the impact that Big Data specialists can have across a range of industries.

By the end of this module, you will have a clear understanding of the opportunities, challenges, and rewards that come with a career in Big Data. You'll have the tools you need to take that first step on your path, and the knowledge to continue growing in your career. We look forward to guiding you on this exciting journey!

Importance of a Career in Big Data

Why should you consider a career in Big Data? Apart from the variety of roles available, there are several compelling reasons.

Job Security and Growth

Firstly, Big Data is an industry experiencing rapid growth. According to a report by IBM, by 2020 there was a 28% increase in demand for data professionals, and the trend has only continued since. This translates into job security, a crucial consideration when selecting a career path.

High Earning Potential

Big Data roles are often well compensated. While the specific salary can vary based on the role, your experience, and location, data roles often command above-average salaries due to the specialized skill set required and the demand for these skills in the market.

Impact Across Industries

A career in Big Data puts you in a position to make significant contributions to any industry, from healthcare to finance, retail to energy. Data-driven insights can lead to breakthroughs, efficiencies, and competitive advantages. For instance, in healthcare, your analysis of patient data could lead to improved treatments or the early detection of diseases.

Personal Growth and Continuous Learning

Finally, Big Data is a field that promotes constant learning and personal growth. The field evolves rapidly, so you'll have ample opportunities to learn, adapt, and upskill. Whether it's mastering a new programming language or getting to grips with the latest machine learning algorithms, your learning curve will always be active.

As you embark on this exciting career journey, remember that your potential is limitless. Your role in Big Data can extend beyond the titles mentioned above. You could become a team lead, a project manager, a consultant, or even a director of data science in a leading organization. With the right mix of skills, experience, and continuous learning, a career in Big Data can open up a world of opportunities for you.

Career Pathways in Big Data

In the realm of Big Data, there is a myriad of career pathways you can choose from, each with its unique set of responsibilities and opportunities. The field of Big Data is highly interdisciplinary and, depending on your interests and skills, you might find one or more of these roles appealing.

Data Analyst

As a Data Analyst, your primary role would be to turn data into information and information into insight. You would work closely with different stakeholders within your organization, helping them make data-driven decisions.

For example, you might be handed a dataset related to your company's recent marketing campaign. Your job would be to analyze this data and answer key questions such as "Did our recent marketing campaign increase sales?" or "What was the return on investment for this campaign?".

Your day-to-day responsibilities would include cleaning and pre-processing data, conducting statistical analyses, creating data visualizations, and preparing reports. Proficiency in SQL and programming languages like Python or R, along with a good understanding of statistical analysis, are crucial skills for a Data Analyst.

Data Scientist

As a Data Scientist, you are expected to delve deeper into the data and extract more complex insights. You would employ a variety of advanced techniques, such as machine learning and predictive modeling, to anticipate future trends or behaviors.

Imagine you are working for an e-commerce company and you have been given access to historical customer transaction data. As a Data Scientist, you could use this data to build a recommendation engine. This engine, powered by machine learning algorithms, could predict what products a customer is likely to buy next, enhancing the overall customer experience and driving additional sales.

Mastering skills such as machine learning, statistical modeling, along with programming languages like Python and R, is key to becoming a successful Data Scientist.

Data Engineer

As a Data Engineer, you would be the backbone of any data-driven operation. Your role involves designing, building, and maintaining the data architecture of your organization.

Consider the scenario where your company is collecting data from various sources, including website logs, transaction databases, and external data feeds. As a Data Engineer, your job would be to build a robust and scalable data pipeline that collects, integrates, and prepares this data for use by Data Analysts and Data Scientists.

Key skills for a Data Engineer include proficiency in SQL, experience with different database systems (both SQL and NoSQL), and knowledge of big data tools and frameworks like Hadoop, Apache Kafka, or Apache Spark.

Business Intelligence Analyst

As a Business Intelligence Analyst, your goal is to help your organization make more strategic decisions. You would analyze data relevant to your business and present actionable insights to executives and other decision-makers.

Suppose your company is looking to expand its product line and needs to understand which products are most likely to be successful. As a Business Intelligence Analyst, you might analyze sales trends, market conditions, and competitor data to recommend which products your company should focus on.

To excel as a Business Intelligence Analyst, you would need to have strong analytical skills, proficiency in SQL, and experience with BI tools like Tableau or PowerBI.

Machine Learning Engineer

As a Machine Learning Engineer, you are at the cutting edge of technology, designing and implementing machine learning models to solve complex business problems.

Imagine you work for a company that develops voice recognition software. As a Machine Learning Engineer, your role could involve developing and refining machine learning models that can accurately transcribe and interpret human speech, continuously improving the accuracy and utility of your company's software.

A Machine Learning Engineer needs to be proficient in a variety of skills, including programming (Python, Java, etc.), machine learning algorithms, and tools and libraries like TensorFlow or PyTorch.

Big Data Architect

As a big data architect, your role would be to design the entire infrastructure for handling your organization's data. This is a strategic role that involves understanding both the technical aspects of big data and the specific business needs of your organization. You would create the blueprint for data management systems to integrate, centralize, protect, and maintain the data sources.

Data Manager

As a data manager, you would oversee all the data assets of the organization. This role involves a wide range of responsibilities such as managing data storage, ensuring data accessibility, maintaining data quality, protecting data privacy, and complying with regulatory standards.

Big Data Project Manager

As a big data project manager, your responsibility would be to oversee big data projects from inception to completion. This role requires strong leadership and organizational skills. You would handle project planning, team coordination, timeline setting, risk management, and ensuring the successful execution of the project goals.

Chief Data Officer (CDO)

As a CDO, you would play a strategic role at the executive level of the organization. You would be responsible for data governance and utilization across the organization. Your key responsibilities would include data protection, privacy, compliance, and ensuring that the organization's data strategy aligns with its business objectives.

Data Governance Manager

As a data governance manager, your role would be to ensure the quality, availability, integrity, security, and usability within an organization's data framework. You would develop and implement data standards, data quality metrics, policies, and related data management practices.

Remember, each of these career pathways offers its unique challenges and rewards, and the choice of the path depends on your interests, skills, and long-term career goals. You might start in one role and transition to another, or specialize in one specific area. There's no fixed path in Big Data – the possibilities are boundless.

Creating an Effective Resume

Now that you have a clear understanding of the various career pathways in Big Data and the skills required, let's focus on one of the most critical aspects of your career journey - creating an effective resume. Your resume is often your first point of contact with potential employers, and it's essential to make a strong impression.

Highlighting Relevant Skills and Experience

Your skills and experiences are the heart of your resume. When writing this section, you need to keep two key points in mind: relevance and clarity.

Start by listing your technical skills - programming languages, databases, machine learning libraries, and any Big Data tools you've used such as Hadoop or Apache Spark. But don't just list them; be specific about how you've used these skills. For instance, instead of writing "Proficient in Python," consider writing "Used Python's Pandas and Matplotlib libraries to clean, analyze, and visualize a dataset of over 1 million records for a personal project."

Next, move on to your experiences. This could include internships, work placements, or even significant academic projects. Like with skills, the key is to focus on what you did and the impact it had. For example, "Developed a machine learning model using TensorFlow to predict customer churn, achieving an accuracy of 85%."

Remember to tailor your resume for each job application. If a job description emphasizes machine learning and Python, make sure those skills are evident in your resume.

Importance of Projects and Portfolios

In the field of Big Data, showing can be even more effective than telling. That's where projects and portfolios come in. A portfolio is a concrete, tangible demonstration of your skills. It provides potential employers with proof of what you can do and what they can expect if they hire you.

If you're just starting out and don't have much work experience, personal projects are a great way to build your portfolio. For instance, you could take a public dataset and conduct a detailed analysis, or build a machine learning model to make predictions.

Include these projects in your resume and, if possible, provide links to where they can be accessed, like a GitHub repository or a personal blog. Explain what you did, the tools and techniques you used, and what the outcomes were. For example, "Built a Python-based web scraper to collect data from several websites, which I then analyzed to identify key market trends. The code and report can be found on my GitHub."

Writing Compelling Cover Letters

Your resume shows what you've done and what you can do, but your cover letter is where you can show who you are. This is where you get to tell your story and express your passion for Big Data.

In your cover letter, you should aim to answer the following questions: Why are you interested in this role? How do your skills and experiences align with the job description? What unique qualities do you bring?

Start by expressing your interest in the role and the company, and why you're attracted to them. Then, align your skills and experiences with the job requirements. Choose 2-3 key requirements and discuss concrete examples that demonstrate you meet these. For example, if the job requires experience with machine learning, you could talk about a project where you successfully built and deployed a machine learning model.

Finally, conclude by mentioning something unique about you that could add value to the team or the company. This could be a soft skill, a unique perspective, or a relevant interest or hobby.

Preparing for the Job Market

With a strong resume and cover letter at your disposal, you're now ready to step into the job market. However, securing a job in Big Data isn't just about having the right skills and a polished resume. It's also about building a strong professional network, acing your interviews, and effectively navigating job offers and salary negotiations.

Networking Strategies

Networking is a powerful tool for career development. It's about creating valuable relationships that can open doors to new opportunities. Here are a few strategies to help you network effectively:

- **Attend Industry Events:** These can range from seminars, webinars, meetups to large conferences. Participating in these events can offer you the chance to learn from industry experts, stay updated on industry trends, and connect with professionals who share your interests.
- **Use Social Media:** LinkedIn is a particularly useful platform for professional networking. Connect with peers, join relevant groups, engage in discussions, and share your thoughts and work. But remember, networking is a two-way street. Don't just reach out to people when you need something; try to provide value as well.
- **Connect with Alumni:** Your school or university alumni network can be a great resource. You can reach out to them for advice, mentorship, or even job opportunities.

Remember, the goal of networking is not just to meet as many people as possible but to build meaningful relationships. Always follow up after meeting someone new, and look for ways to offer help or provide value.

Preparing for Interviews

Interviews are your chance to showcase not only your skills and knowledge but also your passion for Big Data and your cultural fit for the company. Here's how to prepare:

- **Research the Company:** Understand their products or services, their values, culture, and recent news. This shows your interest in the company and can help you tailor your answers.
- **Know the Role:** Read the job description carefully and be prepared to demonstrate how your skills and experiences align with the job requirements.
- **Prepare for Technical Questions:** If you're applying for a technical role, expect questions that test your skills in areas like programming, statistics, machine learning, and Big Data tools.
- **Prepare for Behavioral Questions:** These questions assess your soft skills and cultural fit. Be ready to provide examples that demonstrate qualities like problem-solving, teamwork, and communication.
- **Ask Questions:** Interviews are a two-way street. Prepare thoughtful questions about the role, team, or company. This demonstrates your interest and helps you evaluate if the company is a good fit for you.

Navigating Job Offers and Salary Negotiations

Receiving a job offer can be exciting, but it's important to take some time to evaluate the offer and, if necessary, negotiate.

- **Evaluate the Offer:** Consider all aspects of the offer, including salary, benefits, growth opportunities, the team you'll be working with, the company culture, and how the role aligns with your career goals.

- **Know Your Worth:** Research salary ranges for your role in your location. Websites like Glassdoor can be useful for this. Knowing the market rate will help you understand if the offer is fair.
- **Negotiate, If Necessary:** If the offer is below your expectations or the market rate, consider negotiating. Be polite and respectful, clearly express your concerns, and provide justification for your request, such as your skills, experiences, or the market rate.

Remember, preparing for the job market is a multi-faceted process. With effective networking, thorough interview preparation, and thoughtful navigation of job offers, you'll be well-positioned to land a fulfilling role in Big Data.

Continuous Learning and Development

A career in Big Data is a journey, not a destination. Technologies, tools, and best practices in this field evolve at a rapid pace. Staying relevant requires continuous learning and development. In this section, we will explore how you can stay up-to-date with industry trends, upskill and cross-skill, and leverage conferences, webinars, and workshops for your professional growth.

Staying Up-to-Date with Industry Trends

Being aware of the latest trends in the Big Data field is crucial. It helps you understand where the industry is heading, which tools are gaining popularity, and which skills are most in demand. Here are some practical ways you can stay updated:

- **Follow Industry Blogs and Websites:** Websites like KDnuggets, Towards Data Science, and the Harvard Business Review offer valuable insights into the latest trends and innovations in Big Data.

- **Subscribe to Newsletters:** Newsletters, such as the Data Science Weekly or the O'Reilly Data Newsletter, curate the latest news and articles in the Big Data field.

- **Participate in Online Communities:** Platforms like Stack Overflow, Reddit, or GitHub offer vibrant communities where you can learn from peers, share knowledge, and stay updated on what's new in Big Data.

Upskilling and Cross-Skilling

Upskilling (enhancing your existing skills) and cross-skilling (learning new, related skills) are essential strategies for career growth in Big Data. Here's how you can approach this:

- **Identify the Skills You Need:** Understand the skills in demand for your current role or your desired career path. Job postings, industry reports, and conversations with peers can provide valuable insights.
- **Leverage Online Learning Platforms:** Websites like Coursera, edX, and Udemy offer courses on a wide variety of Big Data topics, from programming and machine learning to data visualization and cloud computing.
- **Work on Projects:** Applying what you learn to real or simulated projects is one of the best ways to cement your new skills. Consider contributing to open-source projects or working on personal projects.

Importance of Conferences, Webinars, and Workshops

Conferences, webinars, and workshops offer unique learning opportunities. They allow you to hear from industry leaders, learn about the latest research and innovations, and network with other Big Data professionals.

- **Conferences:** Big Data conferences like the Strata Data Conference or the International Conference on Big Data offer a mix of keynote presentations, panel discussions, and workshops on a variety of topics. They can be a great place to learn about the latest trends and connect with like-minded professionals.
- **Webinars:** Many organizations and experts in the Big Data field regularly host webinars. These are often more focused and shorter than conferences, making them a convenient way to learn about specific topics.
- **Workshops:** Workshops typically offer hands-on learning experiences. They can be a great way to learn new tools or techniques, often with direct guidance from experts.

Continuous learning is crucial in the Big Data field. By staying up-to-date with industry trends, continuously upskilling and cross-skilling, and actively participating in conferences, webinars, and workshops, you can ensure that you stay relevant and continue to grow in your Big Data career.

Case Studies

A key to navigating your Big Data career journey successfully is to learn from others who have walked the path before you. By studying success stories and understanding the challenges and failures others have faced, you can gain valuable insights to guide your own career.

Success Stories in Big Data Careers

Every successful person in Big Data has a unique story. Studying these stories can provide inspiration, motivation, and practical tips for your own career.

Example 1: The Transitioning Professional

Consider the story of a software developer who successfully transitioned into a Data Scientist role. This person had a strong foundation in programming but knew little about machine learning and statistics. To transition, they took several online courses to build their knowledge of these areas. They also undertook personal projects where they could apply what they had learned, and used these projects to demonstrate their new skills. By combining continuous learning with practical application, they were able to transition into a successful career as a Data Scientist.

Example 2: The Non-Traditional Entrant

Then, there's the story of a professional from a non-technical background, say, humanities, who became a successful Data Analyst. This individual had strong critical thinking skills but lacked the technical skills required for a career in Big Data. Undeterred, they leveraged online learning platforms to learn programming, databases, and data visualization. They also built a portfolio of projects using public datasets, demonstrating that despite their non-traditional background, they had the skills necessary to extract insights from data.

These stories underscore the power of self-learning, the importance of hands-on practice, and the fact that no one's journey in Big Data is identical.

Learning from Failures and Challenges

In any career, challenges and failures are not only inevitable but also valuable learning opportunities. Let's look at some common challenges in Big Data careers and how to learn from them.

Example 1: Difficulty in Keeping Up with New Technologies

One common challenge in Big Data is the fast-paced nature of the field. New tools and technologies emerge regularly, and it can be overwhelming to keep up. In such a scenario, you might not become an expert in every new tool, but understanding the landscape and the purpose of these tools is beneficial. Choose the technologies that are most relevant to your role or your career goals, and focus on learning those deeply.

Example 2: Overcoming the Imposter Syndrome

Imposter syndrome - the feeling that you're not as competent as others perceive you to be - is common in a field as complex and vast as Big Data. If you find yourself experiencing this, know that you're not alone. Everyone, at some point, feels out of their depth. Instead of letting it paralyze you, use it as motivation to learn and improve. Recognize your achievements, seek feedback, and remember that it's okay not to know everything.

Through these case studies, we see that building a successful career in Big Data involves continuous learning, resilience in the face of challenges, and the courage to take on new opportunities. Your journey may be unique, but the underlying principles remain the same. By learning from both the successes and failures of others, you can chart your own path in the exciting field of Big Data.

Conclusion

As we draw this comprehensive guide to a close, it's important to contemplate the future prospects of Big Data careers, the significance of maintaining a healthy work-life balance, and final advice that can guide your journey forward in this exciting field.

Future Prospects of Big Data Careers

Big Data is not just a buzzword; it is a revolution that's here to stay. With data proliferation happening at an unprecedented rate, the importance of professionals who can make sense of this data is increasing exponentially. IDC predicts that by 2025, the global data sphere will grow to 175 zettabytes. This indicates a promising future for Big Data careers.

You're at the right place at the right time. Whether you're a data analyst, data scientist, data engineer, business intelligence analyst, or a machine learning engineer, you will find plenty of opportunities to grow and thrive in your career.

Furthermore, roles in Big Data are not confined to a specific industry. Every industry, from healthcare and finance to entertainment and sports, is leveraging data to make informed decisions. This gives you the flexibility to work in a field that aligns with your interests.

Maintaining Work-Life Balance in a Big Data Career

While pursuing a career in Big Data is rewarding and exciting, it can also be demanding. Balancing professional commitments with personal life can be challenging but is crucial for long-term career success and personal well-being.

Here are some tips to help maintain a healthy work-life balance:

- **Set Boundaries:** It's important to delineate when your workday starts and when it ends. Taking breaks during the day can help rejuvenate your mind and increase productivity.

- **Prioritize Your Tasks:** Not every task needs to be completed right away. Prioritize your tasks based on their importance and deadlines. This can prevent last-minute rushes and reduce stress.

- **Take Care of Your Health:** Regular exercise, a balanced diet, and adequate sleep are non-negotiables. A healthy body is crucial for a healthy mind.

- **Make Time for What You Love:** Be it spending time with your family, pursuing a hobby, or simply relaxing - ensure to make time for activities that make you happy and help you unwind.

Final Advice and Motivation

Building a career in Big Data is a journey, not a destination. Here's some final advice for your journey:

- **Stay Curious and Keep Learning:** Never lose your curiosity. The best data professionals are those who constantly seek to learn and grow.

- **Embrace Challenges:** Challenges are not roadblocks; they are opportunities to learn and improve. Embrace them, learn from them, and come out stronger.

- **Network and Collaborate:** Connect with peers and professionals in the field. Share your work, learn from theirs, and build a supportive community around you.

Remember, the only constant in life is change, especially true in a field as dynamic as Big Data. So, adapt, learn, grow, and most importantly, enjoy the process. The world of Big Data is at your fingertips, ready to be explored and conquered. With hard work, resilience, and the right mindset, you are sure to forge a rewarding and fulfilling career in this vibrant field.

Module 10 Exam Questions

1. What is Big Data characterized by?

a) Large volume

b) Variety of sources

c) High velocity

d) All of the above

2. Which programming language is crucial for big data analysis?

a) JavaScript

b) Python

c) Ruby

d) Swift

3. Which one is not a soft skill for big data careers?

a) Communication

b) Creativity

c) Negotiation

d) Painting

4. What certification could enhance your big data skillset?

a) Certified Big Data Professional

b) Certified Python Programmer

c) Certified Java Developer

d) Certified Web Designer

5. What role involves building and maintaining data pipelines and databases?

a) Data Scientist

b) Data Analyst

c) Data Engineer

d) Business Intelligence Analyst

6. What role involves creating strategic plans from data insights?

a) Data Scientist

b) Data Analyst

c) Data Engineer

d) Business Intelligence Analyst

7. Which role requires a strong understanding of data analysis and software engineering?

a) Machine Learning Engineer

b) Data Analyst

c) Business Intelligence Analyst

d) Big Data Architect

8. Who is responsible for overseeing the data assets of an organization?

a) Big Data Project Manager

b) Data Manager

c) Machine Learning Engineer

d) Data Analyst

9. What does a Big Data Architect do?

a) Design the infrastructure for handling data

b) Manage data projects from start to finish

c) Oversee data assets of an organization

d) Train machine learning models

10. Which role is responsible for managing big data projects from start to finish?

a) Data Manager

b) Big Data Architect

c) Data Scientist

d) Big Data Project Manager

11. Which document allows you to showcase your big data skills and experience?

a) Cover letter

b) Reference letter

c) Personal statement

d) Resume

12. Why is networking important in big data careers?

a) To learn about job openings

b) To stay updated on industry trends

c) To find mentors in your field

d) All of the above

13. What is the purpose of a cover letter?

a) To list your educational qualifications

b) To provide references

c) To highlight your relevant skills and why you're a good fit for the job

d) To provide a full employment history

14. Why is it important to stay updated on the latest Big Data industry trends?

a) To keep your skills relevant

b) To identify new opportunities

c) To anticipate future challenges

d) All of the above

15. Which of the following can help you stay updated on the latest industry trends?

a) Online communities

b) Industry blogs and websites

c) Newsletters

d) All of the above

16. What is Upskilling?

a) Learning new related skills

b) Enhancing your existing skills

c) Forgetting your old skills

d) Ignoring your skills

17. What is a common challenge people face in building a big data career?

a) Feeling overwhelmed by the pace of change

b) Dealing with Imposter Syndrome

c) Maintaining work-life balance

d) All of the above

18. What is the future prospect of big data careers?

a) Decreasing demand

b) Increasing demand

c) No change in demand

d) None of the above

19. What role involves designing models and algorithms for effective data usage?

a) Data Scientist

b) Data Analyst

c) Business Intelligence Analyst

d) Data Engineer

20. Why is work-life balance important in a Big Data career?

a) To avoid burnout

b) To maintain productivity

c) To keep motivation high

d) All of the above

Answers:

1. d
2. b
3. d
4. a
5. c
6. d
7. a
8. b
9. a
10. d
11. d
12. d
13. c
14. d
15. d
16. a
17. d
18. b
19. a
20. d

Conclusions and Next Steps

As we reach the end of this book, it's evident that the Big Data revolution is redefining our world, with its transformative effects palpable across various industries and within our personal lives. Throughout our exploration, we have understood that Big Data is not merely a fleeting trend, but a significant force, powering economic development, enriching our lives, and transforming industries. We've plunged into strategies for effective Big Data management, securing data privacy, and fostering innovation and leadership in Big Data application.

The future will be characterized by the widespread integration of Big Data, revolutionizing industries and making our lives more interconnected than ever before. Recognizing the potentials and challenges of Big Data will allow us to utilize this technology to create more intelligent, secure, and efficient systems and societies. However, this transformation can only truly blossom when each of us actively welcomes the changes instigated by the Big Data era.

As a change-maker in this space, you have the capacity to significantly influence your industry and the wider world. Here are the next steps you can take to embrace the Big Data revolution and drive positive change:

1. **Enroll the online program at TheBestMBA.org:** Certify your skills, gain in-depth knowledge, practical tools, and valuable insights from our comprehensive curriculum. Our program is designed to help you thrive in the Big Data era and adapt to the ever-changing demands of the workforce.

2. **Engage in continuous learning:** Stay ahead of the curve by constantly updating your skills and knowledge. Embrace lifelong learning and take advantage of the vast resources available online, such as courses, webinars, podcasts, and articles.

3. **Expand your network:** Connect with industry leaders, experts, and fellow game changers through Big Data-focused events, conferences, and workshops. Foster meaningful

relationships and collaborations that will propel you forward in your career.

4. **Seek mentorship and guidance:** Reach out to experienced professionals in the field of Big Data and related industries to learn from their insights and experiences. Their wisdom can help you navigate the complexities of the IoT landscape and ensure your success.
5. **Volunteer or collaborate on Big Data projects:** Gain hands-on experience and demonstrate your capabilities by working on real-world IoT projects. This not only enhances your skills but also showcases your value to potential employers and collaborators.

By taking these steps, you will be well on your way to becoming an influential game changer in the Big Data era. Remember, the journey towards a sustainable and inclusive future begins with each individual's commitment to adapt, learn, and grow. Embrace the possibilities that Big Data offers and let your unique human skills shine through as you create a better tomorrow.

The future is now. Are you ready to shape it?

Join the IIEC Big Data Certification Program today available at https://thebestmba.org and become a game changer.

I hope you enjoyed this journey together, and as I mentioned at the beginning of this book, I would love for you to share your experience with me after completing this program. Feel free to message me via LinkedIn.

https://www.linkedin.com/in/angelberniz

To your success,

Angel Berniz

Made in the USA
Monee, IL
04 November 2026

67384020R00174